DEBATING

DARWIN'S DOUBT

DEBATING DARWIN'S DOUBT

A SCIENTIFIC CONTROVERSY
THAT CAN NO LONGER BE DENIED

DAVID KLINGHOFFER, EDITOR

DISCOVERY INSTITUTE PRESS   SEATTLE   2015

Description

This book gathers essays and blog posts responding to criticism of *Darwin's Doubt: The Explosive Origin of Animal Life and the Case for Intelligent Design* by Stephen Meyer. The book explores topics such as orphan genes, cladistics, small shelly fossils, protein evolution, the length of the Cambrian explosion, the God-of-the-Gaps objection to intelligent design, and criticisms raised by proponents of theistic evolution. Contributors include Stephen Meyer, Douglas Axe, David Berlinski, William Dembski, Ann Gauger, Casey Luskin, and Paul Nelson. Edited by David Klinghoffer.

Publisher's Note

This book is part of a series published by the Center for Science & Culture at Discovery Institute in Seattle. Previous books include *Signature of Controversy: Responses to Critics of Signature in the Cell*, edited by David Klinghoffer; *The Myth of Junk DNA* by Jonathan Wells; *The Deniable Darwin & Other Essays* by David Berlinski; and *Discovering Intelligent Design: A Journey into the Scientific Evidence* by Gary Kemper, Hallie Kemper, and Casey Luskin.

Library Cataloging Data

Debating Darwin's Doubt: A Scientific Controversy that Can No Longer Be Denied
Edited by David Klinghoffer.
380 pages, 6 x 9 x 0.79 in. & 1.12 lb., 229 x 152 x 20 cm. & 0.5 kg.
Library of Congress Control Number: 2015941200
BISAC Subject: SCI027000   SCIENCE / Life Sciences / Evolution
BISAC Subject: SCI080000   SCIENCE / Essays
BISAC Subject: SCI034000   SCIENCE / History
ISBN-13: 978-1-936599-28-8 (paperback)

Publisher Information

Discovery Institute Press, 208 Columbia Street, Seattle, WA 98101
Internet: http://www.discoveryinstitutepress.com/
Published in the United States of America on acid-free paper.
First Edition. First Printing. July 2015.

Contents

IV. Biologist: Martin Poenie

V. Miscellaneous Challenges

VI. Trends in Reviewing *Darwin's Doubt*

VII. Replying to *Darwin's Doubt* Without Naming It

VIII. Responses from Theistic Evolutionists

IX. Independent Confirmation of Meyer's Thesis

INTRODUCTION:

No Debate over

Darwinian Evolution?

David Klinghoffer

WHATEVER ELSE STEPHEN MEYER DEMONSTRATED ABOUT THE explosion of biological information required to build the Cambrian animals, his bestseller *Darwin's Doubt* served as a massive rebuke to the mantra-like assertion that there is "no debate," "no controversy" among scientists about Darwinian evolution.

There's plenty! Meyer showed this in the book by addressing the arguments of prominent scientists who seek what they call a "Third Way" (neither intelligent design nor Darwinism) and a new theory of evolution. Nicely coinciding with the release of the paperback edition of *Darwin's Doubt*, these researchers launched a provocative website, *The Third Way*, as a gathering place for those sharing their views.[1] And Meyer showed it again with the new Epilogue, included in the paperback, in which he replied in detail to the most substantive of his critics.

Debating Darwin's Doubt could be thought of as a supplement to what Dr. Meyer wrote in the Epilogue. The reception of *Darwin's Doubt*, with serious scientific thinkers arguing back and forth about his thesis, was definitive evidence that the Darwinist mantra is wrong. If there is no scientific controversy about Darwinism versus intelligent design, how can one explain the volume of disputatious views aired in the wake of the book's release?

Here we have gathered together a sample, collected mostly from writing by Discovery Institute scholars published at our popular news site *Evolution News & Views*. In these pages Stephen Meyer, Douglas Axe, Ann Gauger, William Dembksi, David Berlinski, Casey Luskin, Paul Nelson and I argue with critics including Charles Marshall (UC Berkeley), Robert Asher (Cambridge University), Martin Poenie (University of Texas), Donald Prothero (Natural History Museum of Los Angeles County), Nick Matzke (National Institute for Mathematical and Biological Synthesis), and others.

These critics' arguments in turn were recycled by popular media outlets such as *The New Yorker, National Review, First Things*, and around the Internet including more than six hundred reviews (at the moment) on the book's Amazon page. Nick Matzke's critiques at *Panda's Thumb*, in particular, became a guiding light for the recyclers, brandished again and again by both lofty and anonymous opponents of intelligent design. University of Chicago biologist Jerry Coyne, who writes the blog *Why Evolution Is True*[2] and is arguably the American Richard Dawkins, pointed to Matzke as a justification for failing to wrestle with Meyer's arguments himself. All this accounts for the extensive treatment given here to Matzke.

The critical response to *Darwin's Doubt* was matched by warm appreciation from readers—and from scientists in high places. Harvard University geneticist George Church praised it as "an opportunity for bridge-building, rather than dismissive polarization." Biologist Scott Turner, State University of New York, saw in it "an intriguing exploration of one of the most remarkable periods in the evolutionary history of life."

Russell Carlson, molecular biologist at the University of Georgia, said that it "demonstrates, based on cutting-edge molecular biology, why explaining the origin of animals is now not just a problem of missing fossils, but an even greater engineering problem." Mark McMenamin, paleontologist at Mt. Holyoke College and co-author of *The Emergence of Animals*, greeted the book as "a game changer for the study of evolution."

Bestselling novelist Dean Koontz even threw in an enviable accolade from his own professional perspective: "Meyer writes beautifully. He marshals complex information as well as any writer I've read."

Did I mention that *Darwin's Doubt* hit #7 on the *New York Times* bestseller list, while also making the bestseller ranks at the *Los Angeles Times* and *Publishers Weekly*? So obviously book-buyers were enthusiastic as well.

The challenge in editing this book was one of cutting and sifting from a vast literature in response to *Darwin's Doubt*.

I was sad, for example, to have to exclude everything we wrote about University of Chicago pathologist Stephen Meredith's essay in *First Things*.[3] Meredith used *Darwin's Doubt* as an occasion to launch a bizarre attack on intelligent design as a revival of a medieval religious heresy, occasionalism. That one brought forth a delicious response[4] from *ENV's* Michael Egnor, distinguished pediatric neurosurgeon at SUNY Stony Brook—which, unfortunately, I was unable to include just because Meredith's criticism was hardly scientific at all but, instead, overwhelmingly theological. Strange to say, in an exchange with Stephen Meyer at *The American Spectator*, journalist John Derbyshire likewise brought forward the charge of "occasionalism."[5] At a certain point, criticisms of the theory of intelligent design get so ridiculous and abstracted from reality that you throw up your hands in wonder. Yes, of course many *unserious* thinkers too have joined the mob arrayed against Meyer's book.

It was disappointing not to include Martin Poenie's contribution to *ENV* in the series where he disputed with Biologic Institute's Douglas Axe about the thesis of *Darwin's Doubt*.[6] When I contacted him as a courtesy to ask permission to republish his article, Dr. Poenie first answered that he couldn't remember writing one. When I sent him the link, repeatedly, he then chose not to respond at all. Whether this means Poenie lost confidence in his arguments after having them refuted by Dr. Axe, I cannot tell.

I also would have wanted to include something about theistic evolutionist Karl Giberson's public debate with Stephen Meyer, in which Meyer talked about the evidence for design in the Cambrian explosion. Dr. Giberson, a physicist and theistic evolutionist, displayed a photo of a baby with what appeared to be a "perfectly formed" tail to strengthen his case for our common descent from a tailed ancestor. Later, writing at *The Daily Beast* and once again using the photo, he mused about why Meyer could not refute the argument represented by the image. As we reported at *ENV*, however, the supposed evolutionary vestige turned out to be nothing more than a Photoshopped fake,[7] and Giberson's arguments about human "tails" were betrayed by the medical literature.[8]

Amusing, but if I were to include anything about Giberson, how could I leave out what we wrote about Charles Marshall's much more substantive and interesting debate with Meyer on British radio?[9] And on and on.

Eventually, considerations of page count and affordability decided these questions. Any reason *not* to include an article was welcome. Just considering all the relevant material of our own from *ENV* would have resulted in a book larger than *Darwin's Doubt*, and correspondingly expensive. That was not practical.

So, in any event, here it is, documentary evidence that a lively and illuminating controversy is going on, conducted a high level. If anyone tries to tell you there's "no debate" about Darwinian theory, hand him a copy of *Debating Darwin's Doubt*. That should settle the matter.

Notes

1. *The Third Way*, http://www.thethirdwayofevolution.com.

2. *Why Evolution Is True*, http://whyevolutionistrue.wordpress.com.

3. Stephen Meredith, "Looking for God in All the Wrong Places," *First Things*, February 2014, http://www.firstthings.com/article/2014/02/looking-for-god-in-all-the-wrong-places.

4. Michael Egnor, "Dissecting a Dead Jellyfish: Reading Stephen Meredith on Intelligent Design," *Evolution News & Views*, January 27, 2014, http://www.evolutionnews.org/2014/01/dissecting_a_de081591.html.

5. Stephen B. Meyer and John Derbyshire, "Does Intelligent Design Provide a Plausible Account of Life's Origins?," *The American Spectator*, January-February, 2014, http://

spectator.org/articles/57159/does-intelligent-design-provide-plausible-account-lifes-origins.

6. Martin Poenie, "Douglas Axe, Protein Evolution, and *Darwin's Doubt*: A Reply," *Evolution News & Views*, July 24, 2013, http://www.evolutionnews.org/2013/07/douglas_axe_pro074781.html.

7. David Klinghoffer, "Karl Giberson Apologizes for Photoshopped Image of Tailed Baby," *Evolution News & Views*, June 6, 2014, http://www.evolutionnews.org/2014/06/karl_giberson_a086461.html.

8. Casey Luskin, "Another Icon of Evolution: The Darwinian Myth of Human 'Tails,'" *Center for Science and Culture*, May 22, 2014, http://www.discovery.org/a/23041.

9. Casey Luskin, "A Listener's Guide to the Meyer-Marshall Radio Debate: Focus on the Origin of Information Question," *Evolution News & Views*, December 4, 2013, http://www.evolutionnews.org/2013/12/a_listeners_gui079811.html.

I.
PRE-PUBLICATION
DEBATE & DISCUSSION

*There are no weaknesses in the
theory of evolution.*

EUGENIE SCOTT, NATIONAL CENTER
FOR SCIENCE EDUCATION

Eugenie Scott, quoted in Terrence Stutz, "State
Board of Education debates evolution curriculum,"
Dallas Morning News, January 22, 2009.

1.

Darwinists Waste No Time

William Dembski

In March 2013 at the group blog *Panda's Thumb* (*PT*), University of Washington geneticist Joe Felsenstein posted a piece titled "Stephen Meyer Needs Your Help." In the post, he attempted to disparage Meyer's book *Darwin's Doubt* before its publication by suggesting that right-thinking readers at *PT* kindly contact Meyer now in the hope of redressing the book's likely flaws (to my knowledge, neither Felsenstein nor anyone else at *PT* had an advance copy of Meyer's manuscript). Wrote Felsenstein: "I suggest we help Meyer with his book. These days a book can be revised up until perhaps a month before publication, so there is still time for Meyer to take our advice."[1]

Felsenstein tried to get the ball rolling by offering Meyer the following advice:

> Let me start with my suggestion (but you [i.e., *PT* readers] will have others to add). Dr. Meyer should explain the notion of Complex Specified Information (CSI) and deal carefully with the criticisms of it. Many critics of Intelligent Design argued that it is meaningless. But even those who did not consider it meaningless (and I was one) found fatal flaws in the way Meyer's friend William Dembski used it to argue for ID. Dembski's Law of Conservation of Complex Specified Information was invoked to argue that when we observe adaptation that is much better than could be achieved by pure mutation (monkeys-with-genom-

ic-typewriters), that this must imply that Design is present. But alas, Elsberry and Shallit in 2003[2] found that when Dembski proved his theorem, he violated a condition that he himself had laid down, and I (2007) found[3] another fatal flaw—the scale on which the adaptation is measured (the Specification) is not kept the same throughout Dembski's argument. Keeping it the same destroys this supposed Law. Meyer should explain all this to the reader, and clarify to ID advocates that the LCCSI does not rule out natural selection as the reason why there is nonrandomly good adaptation in nature.[4]

Felsenstein's request for clarification could just as well have been addressed to me, so let me respond, making clear why criticisms by Felsenstein, Shallit, et al. don't hold water.

There are two ways to see this. One would be for me to review my work on complex specified information (CSI), show why the concept is in fact coherent despite the criticisms by Felsenstein and others, indicate how this concept has since been strengthened by being formulated as a precise information measure, argue yet again why it is a reliable indicator of intelligence, show why natural selection faces certain probabilistic hurdles that impose serious limits on its creative potential for actual biological systems (e.g., protein folds, as in the research of Douglas Axe), justify the probability bounds and the Fisherian model of statistical rationality that I use for design inferences, show how CSI as a criterion for detecting design is conceptually equivalent to information in the dual senses of Shannon and Kolmogorov, and finally characterize conservation of information within a standard information-theoretic framework. Much of this I have done in a paper titled "Specification: The Pattern That Signifies Intelligence"[5] (2005) and in the final chapters of *The Design of Life* (2008).[6]

But let's leave aside this direct response (to which neither Felsenstein nor Shallit ever replied). The fact is that conservation of information has since been reconceptualized and significantly expanded in its scope and power through my subsequent joint work with Baylor engineer Robert Marks. Conservation of information, in the form that Felsenstein is still dealing with, is taken from my 2002 book *No Free Lunch*.[7] In 2005,

Marks and I began a research program for developing the concept of conservation of information, and we have since published a number of peer-reviewed papers in the technical literature on this topic (note that Felsenstein published his critique of my work with the National Center for Science Education, essentially in a newsletter format, and that Shallit's 2003 article finally appeared in 2011 with the philosophy of science journal *Synthese*, essentially unchanged in all those intervening years). Here are the two seminal papers on conservation of information that I've written with Robert Marks:

+ "The Search for a Search: Measuring the Information Cost of Higher-Level Search."[8]

+ "Conservation of Information in Search: Measuring the Cost of Success."[9]

For those and other papers that Marks, his students, and I have done to extend the results in these papers, visit the Evolutionary Informatics publications web page.[10]

It follows that another way to see that Felsenstein is blowing smoke is to note that he simply is not up to date on our literature dealing with conservation of information. Moreover, if the example of Jeffrey Shallit is any indicator, that ignorance of our recent work gives the appearance of being willful. In an attempt to engage Shallit on this newer approach to conservation of information, I sent him an email some time back asking for his considered response to it. Here's the email he sent me in reply:

> I already told you—since you have never publicly acknowledged even one of the many errors I have pointed out in your work—I do not intend to waste my time finding more errors in more work of yours.

> I find your failure to acknowledge the errors I have pointed out completely indefensible, both ethically and scientifically.

> Jeffrey Shallit

Actually, I did acknowledge an arithmetic error that Shallit found in my book *No Free Lunch*, though the error itself did not affect my conclusion. But most of what he calls errors have seemed to me confusions in his own thinking. The fact is, Shallit and I were together at a conference in the early 2000s and butted heads there on the question of complex specified information. In this encounter, I was frankly surprised that he could not grasp a crucial yet very basic distinction involving Kolmogorov complexity, namely, that even though it assigns high complexity to incompressible sequences taken individually, it can also assign high complexity to compressible sequences when taken as a subclass within a broader class of sequences.

The premise behind Shallit's email, and one that Felsenstein seems to have taken to heart, is that having seen my earlier work on conservation of information, they need only deal with it (meanwhile misrepresenting it) and can ignore anything I subsequently say or write on the topic. Moreover, if others use my work in this area, Shallit et al. can pretend that they are using my earlier work and can critique them as though that's what they did. Shallit's 2003 paper that Felsenstein cites never got into my newer work on conservation of information with Robert Marks, nor did Felsenstein's 2007 paper for which he desires a response. Both papers key off my 2002 book *No Free Lunch* along with popular spin-offs from that book a year or two later. Nothing else.

So, what is the difference between the earlier work on conservation of information and the later? The earlier work on conservation of information focused on particular events that matched particular patterns (specifications) and that could be assigned probabilities below certain cutoffs. Conservation of information in this sense was logically equivalent to the design detection apparatus that I had first laid out in my book *The Design Inference*.[11]

In the newer approach to conservation of information, the focus is not on drawing design inferences but on understanding search in general and how information facilitates successful search. The focus is therefore not so much on individual probabilities as on probability distributions and how they change as searches incorporate information. My universal probability bound of 1 in 10^{150} (a perennial sticking point for Shallit and Felsenstein) therefore becomes irrelevant in the new form of conservation of information, whereas in the earlier it was essential because there a certain probability threshold had to be attained before conservation of information could be said to apply. The new form is more powerful and conceptually elegant. Rather than lead to a design inference, it shows that accounting for the information required for successful search leads to a regress that only intensifies as one backtracks. It therefore suggests an ultimate source of information, which it can reasonably be argued is a designer. I explain all this in a nontechnical way in an article I posted at *Evolution News & Views* titled "Conservation of Information Made Simple."[12]

So what's the take-home lesson? It is this: Stephen Meyer's grasp of conservation of information is up to date. His 2009 book *Signature in the Cell* devoted several chapters to the research by Marks and me on conservation of information, which in 2009 had been accepted for publication in the technical journals but had yet to be actually published. Consequently, we can expect Meyer's 2013 book *Darwin's Doubt* to show full cognizance of the conservation of information as it exists currently. By contrast, Felsenstein betrays a thoroughgoing ignorance of this literature. Consequently, if Felsenstein is representative of the help that *PT* has to offer the ID community, then Meyer can afford to do without it.

Notes

1. Joel Felsenstein, "Stephen Meyer Needs Your Help," *Panda's Thumb*, March 26, 2013, http://pandasthumb.org/archives/2013/03/stephen-meyer-n.html.

2. Wesley Elsberry and Jeffrey Shallit, "Information Theory, Evolutionary Computation, and Dembski's 'Complex Specified Information,'" November 16, 2003, http://www.talkreason.org/articles/eandsdembski.pdf.

3. Joe Felsenstein, "Has Natural Selection Been Refuted? The Arguments of William Dembski," *National Center for Science Education*, May–August 2007, http://ncse.com/rncse/27/3-4/has-natural-selection-been-refuted-arguments-william-dembski.

4. Felsenstein, "Stephen Meyer Needs."

5. William A. Dembski, "Specification: The Pattern that Signifies Intelligence," August 15, 2005, http://www.designinference.com/documents/2005.06.Specification.pdf.

6. William A. Dembski and Jonathan Wells, *The Design of Life: Discovering Signs of Intelligence in Biological Systems* (Dallas: Foundation for Thought and Ethics, 2008).

7. William A. Dembski, *No Free Lunch: Why Specified Complexity Cannot Be Purchased Without Intelligence* (Lanham, MD: Rowman & Littlefield, 2002).

8. William A Dembski and Robert J. Marks II, "The Search for a Search: Measuring the Information Cost of Higher-Level Search," *Journal of Advanced Computational Intelligence and Intelligent Informatics* 14, no. 5 (2010): 475–486. Available at: http:// www.evoinfo.org/publications/search-for-a-search/.

9. William A Dembski and Robert J. Marks II, "Conservation of Information in Search: Measuring the Cost of Success," *IEEE Transactions on Systems, Man and Cybernetics A, Systems & Humans* 5, no. 5 (September 2009): 1051–1061. Available at: http://www. evoinfo.org/publications/cost-of-success-in-search/.

10. Found at: http://www.evoinfo.org/publications/.

11. William A. Dembski, *The Design Inference: Eliminating Chance through Small Probabilities* (New York: Cambridge University Press, 1998).

12. William A. Dembski, "Conservation of Information Made Simple," *Evolution News & Views*, August 28, 2012, http://www.evolutionnews.org/2012/08/conservation_ of063671.html.

2.

GRIPE-FEST TURNS SURREAL

David Klinghoffer

A T A CERTAIN POINT, THE PREPUBLICATION GRIPE-FEST BY DARwinian biologist-bloggers about *Darwin's Doubt* turned surreal.

Prophylactically, Jerry Coyne and Joe Felsenstein tried to ward off evident anxieties about Stephen Meyer's book by assuring fellow Darwinists they knew what was in it, and then attacking it on those grounds.[1] Larry Moran, who of course also hadn't read it, endorsed Dr. Coyne's delusional summation[2] of the book's contents ("Yes, baby Jesus made the phyla!") and went after Casey Luskin for the *ethical violation*, no less, of writing about the book prepublication—though of course Casey *had* read it.

> The Intelligent Design Creationists want you to know that any criticism of what they are saying about the book is unethical unless you've read it yourself. However, it's not the least bit unethical for them to make outlandish claims about what's in the book months before we can verify whether those claims are correct.

> This is creationist ethics. It's not supposed to make sense.[3]

Moran took off after me too for "speculat[ing] about what the book is going to say":

> Don't make outlandish claims about what's in a book until it's published and everyone can check for themselves. If you specu-

late about what the book is going to say then don't be surprised if others do as well.[4]

But I was not speculating—at that point I had read *Darwin's Doubt* too. I had the unbound galley right in front of me on my desk. Moran promised to read the book, though he complained that despite the June 18 publication date he likely wouldn't be able to get hold of a copy till August since he lives in Canada.

Canada? Not Timbuktu. It takes two months to ship a book to Toronto? That is very weird.

Anyway, let's get the Moran logic clear. It's perfectly OK to review a book you haven't read before it's published, if the book argues for intelligent design and you are attacking it in absurd cartoon terms as Coyne did. But writing about the same book before it's published, if you *have read it* and are favorably impressed by its argument, is an ethical breach. You see, this is really how these guys think.

Notes

1. David Klinghoffer, "Current Trends in Darwinian Book Reviewing," *Evolution News & Views*, April 17, 2013, http://www.evolutionnews.org/2013/04/current_trends071321.html.

2. Lawrence A. Moran, "Darwin Doubters Want to Have Their Cake and Eat it too," *Sandwalk*, April 18, 2013, http://sandwalk.blogspot.com/2013/04/darwin-doubters-want-to-have-their-cake.html.

3. Lawrence A. Moran, "Soon to Be Released: Another Landmark for the ID Movement," *Sandwalk*, April 12, 2013, http://sandwalk.blogspot.com/2013/04/soon-to-be-released-another-landmark.html.

4. Moran, "Darwin Doubters."

II.
SPEED READER:
NICK MATZKE

Scientists can treat evolution by natural selection as, in effect, an established fact.

NATURE MAGAZINE

"Announcement: Evolutionary gems," *Nature*, 457 (January 1, 2009): 8, http://www.nature.com/nature/journal/v457/n7225/pdf/457008b.pdf.

3.

RUSH TO JUDGMENT

Casey Luskin

THERE'S AN OLD JOKE ABOUT A BOOK CRITIC. ASKED IF HE HAS read the new book by a certain author, he replies, "No, I only had time to review it."

On Wednesday, June 19, 2013, the day after Stephen C. Meyer's book *Darwin's Doubt* was published and made available for purchase, Nick Matzke, at the time a UC Berkeley grad student and now a Postdoctoral Fellow at the National Institute for Mathematical and Biological Synthesis, posted a harsh 9400+ word review on the blog *Panda's Thumb*.[1]

Subsequently, University of Chicago evolutionary biologist Jerry Coyne, perhaps the most prominent American advocate of neo-Darwinism, touted the review as an "excellent" critique.[2] Because of Coyne's prominence, and his endorsement of Matzke's review, it's worth evaluating a few of Matzke's claims.

Now, *Darwin's Doubt* runs to 413 pages, excluding endnotes and bibliography. Neither the book's publisher, HarperOne, nor its author sent Matzke a prepublication review copy. Did Matzke in fact read its 400+ pages and then write his 9400+ word response—roughly 30 double-spaced pages—in little more than a day?

Perhaps, but a more likely hypothesis is that he wrote the lion's share
of the review before the book was released based upon what he presumed
it would say. A reviewer who did receive a prepublication copy, University of Pittsburgh physicist David Snoke, writes:

> A caution: this is a tome that took me two weeks to go through in
> evening reading, and I am familiar with the field. Like the classic
> tome *Gödel, Escher, Bach*, it simply can't be gone through quickly.
> I was struck that the week it was released, within one day of ship-
> ping, there were already hostile reviews up on Amazon. Simply
> impossible that they could have read this book in one night.[3]

Even if Snoke is wrong, and Matzke possesses a preternatural capacity to read and write at blinding speed, Matzke in his haste made
some significant errors—of commission and omission—in his representation and assessment of Meyer's work.

Matzke misrepresents what Meyer actually says, going so far as
to attribute quotes and arguments to him that nowhere appear in the
book. He also fails to address, let alone to refute, Meyer's central arguments. Instead, he attempts to impugn Meyer's credibility by asserting
that Meyer makes various minor factual errors, which turn out not to be
errors at all. Most unfortunately, Matzke gets personal, asserting that
these purported mistakes show that Meyer is ignorant, lazy, arrogant,
and even dishonest. Matzke writes:

> Here it is completely clear that the creationists/IDists are arro-
> gant enough to call God down from Heaven to cover for their ig-
> norance, basically because they are unwilling to do the basic "due
> diligence" and hard work required to get a basic understanding of
> the topic they [sic] commenting on. I'm not sure that most long-
> lived religious traditions actually support that kind of behavior.[4]

And so Matzke attempts to convince readers that they should dis-
trust the man, Stephen Meyer, and ultimately disregard the book that he
has authored, a strategy that Matzke and his colleagues at the National
Center for Science Education have repeatedly used to suppress interest in and consideration of the evidence for intelligent design. Thus, the
punch line of Matzke's review: "I'm not sure it [*Darwin's Doubt*] deserves
much more of anyone's time."[5]

Since I think it would be a shame for readers to miss out on what Meyer has actually written, and since Matzke and other early reviewers on Amazon grossly misrepresent Meyer's argument, I want to set the record straight.

In *Darwin's Doubt*, Meyer argues that the Cambrian explosion presents two separate challenges to contemporary neo-Darwinian evolutionary theory—the first of which Darwin himself also acknowledged in 1859 as a problem for his original theory of evolution. Meyer argues that the geologically sudden appearance of many novel forms of animal life in the Cambrian period, and the absence of fossilized ancestral precursors for most of these animals in lower Precambrian strata, challenges the gradualistic picture of evolution envisioned by both Darwin and modern neo-Darwinian scientists—a problem that many paleontologists have long acknowledged.

More importantly, Meyer argues that the neo-Darwinian mechanism lacks the creative power to produce the new animal forms that first appear in the Cambrian period, a view that many evolutionary biologists themselves also now share. Meyer, in particular, argues that the mutation and natural selection mechanism lacks the creative power to produce both the genetic and epigenetic information necessary to build the animals that arise in the Cambrian explosion. Meyer offers five separate lines of evidence and arguments to support this latter claim. He also later describes and critiques six post neo-Darwinian evolutionary theories and makes a positive argument for intelligent design as well.

Matzke does attempt to address the first problem posed by the Cambrian explosion. He does so by claiming that methods of phylogenetic reconstruction can establish the existence of Precambrian ancestral and intermediate forms—an unfolding of animal complexity that the fossil record does not document. Though he accuses Meyer of being ignorant of these phylogenetic methods and studies, he seems unaware that Meyer in his fifth and sixth chapters explains and critiques attempts to reconstruct phylogenetic trees based upon the comparisons of anatomical and genetic characters. He also criticizes Meyer for being ignorant of cladistics in reconstructing such phylogenetic trees, though, again, Meyer critiques many of the assumptions and methods of cladistics in the context of the larger evaluation of phylogenetic reconstruction that he (Meyer) offers in those chapters (as well as in accompanying endnotes, as I'll explain).

One could say more in response to Matzke's substantive claims about phylogenetic analysis. For now, I recommend Meyer's book itself, specifically his chapters titled "The Genes Tell the Story?" and "The Animal Tree of Life," for any interested reader wanting to know about the problems (that Matzke does not report) with reconstructing phylogenetic trees using these methods. Though Matzke gives the impression of having dealt with these chapters in his review, he doesn't.

Indeed, Matzke scarcely addresses Meyer's second and more central critique of neo-Darwinism. Matzke does not respond in any detail to any of Meyer's multiple challenges to the creative power of the mutation/selection mechanism.

He does not attempt to show that the neo-Darwinian mechanism can efficiently search combinatorial sequence space or attempt to refute empirical studies showing that functional genes and proteins are exceedingly rare within such spaces.

Nor does he show that the mechanism can generate multiple co-ordinated mutations within realistic waiting times—except to reassure us without justification that the need for such mutations is exceedingly rare.

Nor does he explain how the neo-Darwinian mechanism could ever produce new body plans given that mutagenesis experiments show how early-acting body-plan mutations—the very mutations that would be necessary to produce whole new animals from a pre-existing animal body plan—inevitably produce embryonic lethals.

He does not address Meyer's critique of the neo-Darwinian mechanism by explaining how mutations could alter development gene regulatory networks to produce new developmental regulatory networks, though the production of such a new regulatory network is an important requirement for building any new animal body plan from a pre-existing body plan.

Finally, Matzke does not explain how mutations in DNA alone could produce the epigenetic ("beyond the gene") information necessary to build new animal body plans, a problem that has led many evolutionary biologists to seek a new theory of and mechanism for major evolutionary innovation.

Of course, he also fails to show how any of the "post-Darwinian" models critiqued by Meyer could have produced the requisite information for generating animal complexity.

Meyer offers five detailed scientific critiques of the alleged creative power of the mutation/selection mechanism. Yet Matzke in his nearly 10,000-word review offers no detailed response to any. Since Matzke does not address the central critical arguments of Meyer's book, he has not refuted them or shown, therefore, that *Darwin's Doubt* lacks scientific merit.

So what *does* Matzke do in his review?

Apart from arguing that phylogenetic analysis circumvents the problem of missing ancestral fossils, Matzke mainly attempts to impeach Meyer's credibility by pointing to minor, alleged factual errors— errors that, even if Meyer had committed them, would not matter to the substance of the book. Nevertheless, even here Matzke's review fails because Meyer either does not make the errors that Matzke claims, or the "errors" he alleges are not in fact errors.

Matzke claims that Meyer makes two clear-cut "errors" in *Darwin's Doubt* pertaining to proper schemes of taxonomic classification. He alleges, first, that Meyer blunders by calling *Anomalocaris* (literally "abnormal shrimp") an "arthropod." And he claims, second, that Meyer incorrectly calls Lobopodia a phylum.

In his Amazon review, Matzke writes:

> He makes basic errors like calling *Anomalocaris* an arthropod and calling lobopods a "phylum," not noting for the readers that *Anomalocaris* falls well outside of the crown arthropod phylum,

far down in the lobopods, and that the phylum Arthropoda is thought to have evolved from lobopods, as did one or two other phyla. The lobopods are a paraphyletic assemblage of stem taxa, i.e. the very "transitional forms" between phyla that Stephen Meyer claims to be looking for! This is Cambrian Explosion 101 stuff that Meyer gets wrong.[6]

Matzke says much the same on *Panda's Thumb*:

> Meyer continually and blithely refers to organisms such as *Anomalocaris* as "arthropods," as if this were an obvious and uncontroversial thing to say. But in fact, anyone actually mildly familiar with modern cladistic work on arthropods and their relatives would realize that *Anomalocaris* falls many branches and many character steps below the arthropod crown group (see the figure above). *Anomalocaris* lacks many of the features found in arthropods living today. It is one of many fossils with transitional morphology between the crown-group arthropod phylum, and the next closest living crown group, Onychophora (velvet worms).[7]

Notably, these supposed errors pertain to classification, which is a *highly subjective* science. There are strong differences of opinion and much debate about many points regarding the proper classification of Cambrian animals. In fact, Matzke effectively concedes this point in his review, as he calls the definition of phyla "arbitrary and flexible," thus undercutting his own accusations. What Matzke calls "basic errors" really just reflect differences of opinion among experts about how to best classify different Cambrian animals. Even so, Stephen Meyer cites prominent authorities in support of his judgments and positions on classification. Yet in criticizing Meyer, Matzke chooses to ignore that supporting technical literature.

Oddly, Matzke also offers no page numbers for his claim that Meyer calls *Anomalocaris* an arthropod, but I'm happy to fill readers in about what Meyer actually wrote. Meyer addresses this topic on pages 53 and 60 of *Darwin's Doubt*, writing:

> *Anomalocaris* (literally, "abnormal shrimp") and *Marrella*... had hard exoskeletons and clearly represent either arthropods or creatures closely related to them. Yet each of these animals possessed many distinct anatomical parts and exemplified different

ways of organizing these parts, thus clearly distinguishing themselves from better-known arthropods such as the previous staple of Cambrian paleontological studies, the trilobite.[8]

There are many types of arthropods that arise suddenly in the Cambrian—trilobites, *Marrella*, *Fuxianhuia protensa*, *Waptia*, *Anomalocaris*—and all of these animals had hard exoskeletons or body parts.[9]

In the first quote, from page 53, we see that Meyer called *Anomalocaris* "either arthropods or creatures closely related to them," showing his awareness that there is ambiguity and debate over whether *Anomalocaris* belongs directly within arthropods, or was a close relative. Matzke never quotes Meyer's statement on this point, which is consistent both with what Matzke says about anomalocaridids, and with the relevant scientific literature. Instead, Matzke seems unfamiliar with what Meyer wrote.

In the second quote, from page 60, Meyer suggests that *Anomalocaris* may in fact be an arthropod. Would it be a "basic error" to make that claim? Not at all, because many leading authorities on the Cambrian explosion have suggested precisely the same thing —that *Anomalocaris* **is** an arthropod.

One authoritative source on this point is a 2011 *Nature* paper about anomalocaridids by Paterson et al., titled "Acute vision in the giant Cambrian predator *Anomalocaris* and the origin of compound eyes," which concludes:

> These fossils also provide compelling evidence for the arthropod affinities of anomalocaridids, [and] push the origin of compound eyes deeper down the arthropod stem lineage.[10]

The paper firmly places anomalocaridids as stem-group arthropods, very close to the crown-group arthropods, and has some weighty co-authors, including John R. Paterson of the University of New England in Australia, Diego C. García-Bellido of the Instituto de Geociencias in Spain, Michael S. Y. Lee of South Australian Museum and the University of Adelaide, Glenn A. Brock of Macquarie University, James B. Jago of the University of South Australia, and Gregory D. Edgecombe of the Natural History Museum in London. In covering this paper, *Dis-*

cover Magazine stated: "Paterson also argues that the eyes confirm that Anomalocaris was an early arthropod, for this is the only group with compound eyes."[11]

Likewise Benjamin Waggoner (then of UC Berkeley, now at the University of Central Arkansas) writes in the journal *Systematic Biology* that "the anomalocarids and their relatives (Anomalopoda) fall out very close to the base of the traditional Arthropoda and should be included within it."[12] A 2006 paper in the journal *Acta Palaeontologica Polonica* likewise refers to the "anomalocaridid arthropods."[13] The authoritative book *The Cambrian Fossils of Chengjiang China* states that anomalocaridid "morphology recalls features of several phyla, including worm groups, lobopodians, and arthropods. They have been regarded as related to one of these groups, or as arthropods, or as forming an unrelated group."[14] The leading authorities Charles R. Marshall and James W. Valentine note in a 2010 article in the journal *Evolution*, titled "The importance of preadapted genomes in the origin of the animal bodyplans and the Cambrian explosion,"[15] that "*Anomalocaris* most likely lies in the diagnosable stem group of the Euarthropoda (but in the crown group of Panarthropoda)." Writing in *Biological Reviews of the Cambridge Philosophical Society*, Graham Budd and Sören Jensen call *Anomalocaris*, along with *Opabinia*, "stem group arthropods."[16] In the journal *Integrative & Comparative Biology*, Nicholas Butterfield of the University of Cambridge calls *Anomalocaris* an "arthropod":

> A number of arthropods, however, also feature conspicuously three-dimensional phosphatized gut structures, most notably *Leanchoilia* (Fig. 4), *Odaraia*, *Canadaspis*, *Perspicaris*, *Sydneyia*, *Anomalocaris* and *Opabinia*.[17]

Meyer doesn't try to enter into the debate over whether *Anomalocaris* is a "stem group" or "crown group" arthropod, or a member of Euarthropoda, or Panarthropoda. But in calling it an "arthropod" of some type, Meyer can cite many, many scientific authorities who agree with his judgment. Since Meyer states that anomalocaridids are "either arthropods or creatures closely related to them," and that they "possessed many distinct anatomical parts and exemplified different ways of organizing these parts, thus clearly distinguishing themselves from better-known arthropods," his position is smack dab in the middle of the consensus view.[18]

But, of course Matzke doesn't accuse *Nature*, Budd, Jensen, or the authors of any of these other papers of committing a "basic error" by calling *Anomalocaris* an "arthropod." In fact, it seems that Matzke is the one who made the "basic error" in not reading Meyer carefully, not knowing what the literature says, or both.

Now in checking out Matzke's claim, we discovered that there is an actual error in *Darwin's Doubt* regarding *Anomalocaris*, though it isn't anything that Matzke caught. On page 53 Meyer says that *Anomalocaris* has an exoskeleton—when it would have been more accurate to state things as he did on page 60, where he said the anomalocarids "had hard exoskeletons or body parts." The jaw of anomalocaridids is commonly preserved as a fossil, probably because it is a hard part, which was used to hunt hard-shelled organisms. So *Anomalocaris* probably did have "hard parts," but it did not have an exoskeleton. Meyer will correct the oversight in a future edition and note it on his website.

And what about Matzke's other accusation of an alleged error—his claim that Lobopodia isn't a phylum? Again, we're talking about questions of classification here, where there have been many disagreements among scientists about just what is, or isn't, a phylum. Recall that Matzke himself called phyla "arbitrary and flexible." True, there are paleontologists who don't consider Lobopodia a phylum. Some call it a "superphylum," others just a "taxon"; some say it's paraphyletic and others that it is monophyletic. But there are weighty authorities that do consider Lobopodia a phylum.

For example, Lobopodia has been called a "phylum" by one of the leading Cambrian paleontologists, Douglas Erwin, and his co-authors on a 2011 paper in *Science*, "The Cambrian Conundrum: Early Divergence and Later Ecological Success in the Early History of Animals."[19] Their paper attempted an ambitious comprehensive survey of the first appearance of all of the animal phyla in the fossil record, and in a table of supplemental data, they list "Lobopodia" as a "phylum" that first appears in the Cambrian.[20] The table also contains a separate listing for phylum Tardigrada, showing that Meyer was justified in following Erwin et al. by listing the two groups as separate phyla, in contrast to Matzke's charge of "huge mistakes" on this point. Note also that this same table is presented and endorsed by Douglas Erwin and James Valentine in their authoritative 2013 book, *The Cambrian Explosion*.[21]

Other authorities confirm the point about Lobopodia. In a paper in *Biological Review of the Cambridge Philosophical Society*, the eminent Thomas Cavalier-Smith has also classified Lobopodia as one of "three new phyla" in the animal kingdom.[22] The respected biology website Palaeos.com likewise suggests it could be a phylum.[23]

But perhaps the most authoritative source on this is the book *The Cambrian Fossils of Chengjiang, China: The Flowering of Early Animal Life*. Chapter 14 bears the title: "Phylum Lobopodia"![24] This pretty much puts to rest any claims by Matzke that Meyer made some kind of a "basic error" by referring to phylum Lobopodia.

As a side note, Erwin et al.'s 2011 paper (as well as some other leading authorities) also challenges Matzke's claim that the Cambrian explosion lasted 30+ million years.[25] Let it suffice to say that both of Matzke's supposed examples of Meyer's "basic errors" turn out to be bogus.

Again, even if Meyer had made errors in classifying one of the many Cambrian animals, that would in no way affect the strength of his overall argument. The Cambrian fossil record would pose the same two problems to neo-Darwinian theory that Meyer describes at length in his book. Matzke is simply nitpicking.

Matzke claims that Meyer is ignorant of basic concepts in systematics and evolutionary classification, like "stem groups" and "crown groups." He also implies that Meyer doesn't understand why some systematists today reject attempts to classify organisms within traditional Linnean categories such as "phyla." Nevertheless, Meyer provides ample discussion of topics like rank-free classification systems, as anyone who read the book carefully would know. Page 419 of *Darwin's Doubt* has a very nice discussion of stem groups and crown groups, and on pages 31–33, 43, 55, and 418–419, Meyer writes about the concept of "rank-free classification," currently fashionable among some systematists, as well as other phylogenetic concepts.

Meyer's book isn't intended to be a treatise on classification, so he doesn't devote pages and pages to these controversial and hotly debated topics, but Meyer certainly devotes significant space to topics that Matzke claims that he ignores (or doesn't understand). As I noted earlier, Chapters 5 and 6 discuss many fundamental assumptions in phylogenetic methods and critique them, demonstrating a familiarity with the methods and the literature. Matzke utterly fails to engage this discussion—the word "assumption" isn't even in his review. Despite Matzke's many words on phylogenetics, he never offers any evidence that Meyer is ignorant of these topics and concepts or their relevance to his discussion of the Cambrian explosion. Nonetheless, Matzke makes bizarre charges like this:

> I think that if you plunked those fossils down in front of an ID advocate without any prior knowledge except the general notion of taxonomic ranks, the ID advocate would place most of them in a single family of invertebrates, despite the fact that phylogenetic classification puts some of them inside the arthropod phylum and some of them outside of it.[266]

In another instance, Matzke enthusiastically mentions concepts in phylogenetic tree construction like long branch attraction, hoping they can resolve the many conflicts among phylogenetic trees. He criticizes Meyer for neglecting to consider these proposals. But Matzke seems unaware that Meyer has a lengthy (450+ words) endnote on page 432 where he not only writes about long branch attraction, but explains why that idea and many other *ad hoc* explanations fail to account for conflicts among phylogenetic trees. Though Matzke accuses Meyer of ignorance of these phylogenetic concepts and proposals, it rather appears that Matzke is ignorant of Meyer's discussion of them.

Indeed, Matzke's review betrays scant familiarity with the substance of Meyer's book. Of the 9400+ words in the review, fewer than

150 words are actual quotations from *Darwin's Doubt*. In fact, of the
30 or so apparent quotes in his post, all but four are *three words or less*.
For example, Matzke seems to have presciently anticipated that Meyer
would use terms such as "intelligent design," "explosion," "information,"
"phylum," and "fish."

In fact, I find only four instances where Matzke quotes Meyer at a
length of more than five words at a time, which together total about 116
words from the book.

While I personally tend to suspect he didn't read *Darwin's Doubt*,
in the end it doesn't matter whether Matzke read it before writing the
bulk of his review, or whether he wrote it before the book was out based
on presuppositions and then glanced through the pages once he had it in
hand. Either way, his misrepresentations in matters large and small are
inexcusable.

For example, he attributes to Meyer, in quotation marks, phrases
that nowhere appear in the book. Matzke claims Meyer used the phrases
"ancestral phyla," "multiple mutations required," or "conflicts between
trees," but I cannot find those phrases in the book. Some of these mis-
quotes (like "multiple mutations required" or "conflicts between trees")
aren't necessarily far off from things Meyer does say, but Matzke's invention of quotations, like the bogus accusations of "errors," impeaches his
credibility as a reviewer of Meyer's work.

But there are obvious examples of invented quotes, such as when
he claims Meyer argues that "poof, God did it"—words he attributes
to Meyer by putting them in quotation marks, though obviously Meyer
never argued or said such a thing. Matzke offers other descriptions of
Meyer's argument for design which bear no resemblance to Meyer's extensive, rigorous explication of the positive case for design in Chapters
17 through 20. You'd never know it from reading Matzke's review, but
Meyer explains why the standard scientific methods of historical sciences and rigorous abductive logic establish intelligent design as the only
known sufficient cause for generating the information and top-down de-

sign that are required to build the animal body plans that appear explosively in the Cambrian period.

To cite another example, Matzke claims that Meyer said the Cambrian explosion was "instantaneous." Actually, the word "instantaneous" does appear in *Darwin's Doubt* in two places—but *neither is from prose originally written by Meyer,* and neither is necessarily intended to specifically describe the Cambrian explosion. In fact, both uses of that word in the book are from quotes by Stephen Jay Gould. Here's what Gould says:

> But the earth scorns our simplifications, and becomes much more interesting in its derision. The history of life is not a continuum of development, but a record punctuated by brief, sometimes geologically instantaneous, episodes of mass extinction and subsequent diversification.[277]

> Most evolutionary change, we argued, is concentrated in rapid (often geologically instantaneous) events of speciation in small, peripherally isolated populations (the theory of allopatric speciation).[288]

Yet Matzke attributes the word "instantaneous" to Meyer, and calls Meyer "ignorant" because (in his view) "[t]he 'body plans' did not originate instantaneously." But I can find nowhere in the book where Meyer uses the word "instantaneous" to describe the origin of body plans. Needless to say, Matzke doesn't tell us where Meyer allegedly says this—because he never provides pages numbers for the quotes he attributes to Meyer.

In the end it's hard to reconcile Matzke's tone of intellectual superiority with his sloppy scholarship. But, of course, the issue of real interest here is not Nick Matzke. It is the book he supposedly reviewed and refuted, but in fact did not. And since he did not respond to the central arguments of the book that Stephen Meyer actually wrote, readers curious about *Darwin's Doubt* would do well to ignore Matzke's advice. Get the book, and by all means, spend some time with it and read it for yourself.

Notes

1. Nick Matzke, "Meyer's Hopeless Monster Part II," *Panda's Thumb*, June 19, 2013, http://pandasthumb.org/archives/2013/06/meyers-hopeless-2.html.

2. Jerry Coyne, "The First Review of Stephen Meyer's New ID Book," *Why Evolution is True*, June 21, 2013, http://whyevolutionistrue.wordpress.com/2013/06/21/the-first-review-of-stephen-meyers-new-id-book/.

3. David Snoke, "Review of Steve Meyer's New Book, 'Darwin's Doubt,'" *The Christian Scientific Society*, June 21, 2013, http://www.christianscientific.org/review-of-steve-meyers-new-book-darwins-doubt/.

4. Matzke, "Hopeless Monster Part II."

5. Ibid.

6. Ibid.

7. Ibid.

8. Stephen C. Meyer, *Darwin's Doubt: The Explosive Origin of Animal Life and the Case for Intelligent Design* (New York: HarperOne, 2013), 53.

9. Meyer, *Darwin's Doubt*, 60.

10. John R. Paterson et al. "Acute vision in the giant Cambrian predator Anomalocaris and the origin of compound eyes," *Nature* 480, no. 7376 (December 8, 2011): 237–40, http://www.nature.com/nature/journal/v480/n7376/full/nature10689.html.

11. Ed Young, "The sharp eyes of Anomalocaris, a top predator that lived half a billion years ago," *Discover*, December 7, 2011, http://blogs.discovermagazine.com/notrocketscience/2011/12/07/anomalocaris-sharp-eyes-predator/.

12. Benjamin M. Waggoner, "Phylogenetic Hypotheses of the Relationships of Arthropods to Precambrian and Cambrian Problematic Fossil Taxa," *Systematic Biology* 45, no. 2 (1996): 190–222, http://sysbio.oxfordjournals.org/content/45/2/190.abstract.

13. Jianni Liu et al., "A large xenusiid lobopod with complex appendages from the Lower Cambrian Chengjiang Lagerstätte," *Acta Palaeontologica Polonica* 51, no. 2 (2006): 215–222, http://www.app.pan.pl/article/item/app51-215.html.

14. Xianguang Hou, *The Cambrian Fossils of Chengjiang, China: The Flowering of Early Animal Life* (Malden, MA: Blackwell, 2004), 94.

15. C. R. Marshall and J. W. Valentine, "The importance of preadapted genomes in the origin of the animal bodyplans and the Cambrian explosion," *Evolution* 64, no. 5 (May 2010): 1189–0201, http://www.ncbi.nlm.nih.gov/pubmed/19930449.

16. G. E. Budd and S. Jensen, "A critical reappraisal of the fossil record of the bilaterian phyla," *Biol Rev Camb Philos Soc* 75, no. 2 (May 2000): 253–95, http://www.ncbi.nlm.nih.gov/pubmed/10881389.

17. Nicholas J. Butterfield, "Exceptional Fossil Preservation and the Cambrian Explosion," *Integrative and Comparative Biology* 43, no. 1 (2003): 166–177, http://icb.oxfordjournals.org/content/43/1/166.abstract.

18. Meyer, *Darwin's Doubt*, 53.

19. Douglas Erwin et al., "The Cambrian Conundrum: Early Divergence and Later Ecological Success in the Early History of Animals," *Science* 334, no 6059 (November 25, 2011), 1091–1097, http://www.sciencemag.org/content/334/6059/1091.full.html.

20. Ibid.; see supporting online material at, http://www.sciencemag.org/content/suppl/2011/11/22/334.6059.1091.DC1/Erwin.SOM.pdf.

21. Douglas Erwin and James Valentine, *The Cambrian Explosion: The Construction of Animal Biodiversity* (Greenwood Village, CO: Roberts and Company, 2013), 350.

22. T. Cavalier-Smith, "A revised six-kingdom system of life," *Biological Reviews of the Cambridge Philosophical Society* 73, no. 3 (August 1998): 203-66, http://onlinelibrary.wiley.com/doi/10.1111/j.1469-185X.1998. tb00030.x/abstract.

23. "Panarthropoda: Lobopodia," *Palaeos* (July 15, 2002), http://palaeos.com/metazoa/ecdysozoa/panarthropoda/lobopodia.html.

24. Xianguang Hou, *The Cambrian Fossils of Chengjiang, China: The Flowering of Early Animal Life* (Malden, MA: Blackwell, 2004), 82.

25. See *Debating Darwin's Doubt*, Chapter 6, "How 'Sudden' Was the Cambrian Explosion?"

26. Matzke, "Meyer's Hopeless Monster Part II."

27. Stephen Jay Gould, *Wonderful Life: The Burgess Shale and the Nature of History* (New York: Norton, 1990), 54. Quoted on pages 15–16 of *Darwin's Doubt*.

28. Stephen Jay Gould and Niles Eldredge, "Punctuated Equilibria: The Tempo and Mode of Evolution Reconsidered," *Paleobiology* 3 (1977): 116–17. Quoted in *Darwin's Doubt*, chapter 7, endnote 29 on page 434 in the hardback edition and page 469 in the paperback edition.

4.

Matzke, Cladistics, and Missing Ancestors

Stephen C. Meyer

O F THE REVIEWS OF *DARWIN'S DOUBT*, ONE IN A SEEMINGLY OUT-of-the-way venue emerged as a touchstone for many others. Again and again, writers in journals ranging from *The New Yorker* to the ecumenical monthly *First Things* cited a review by Nicholas Matzke that appeared on *Panda's Thumb*, a popular blog dedicated to defending evolutionary theory. University of Chicago evolutionary biologist Jerry Coyne, author of the widely read website *Why Evolution Is True*, has emerged in recent years as an American equivalent of Richard Dawkins, the popular proselytizing spokesman for the neo-Darwinian viewpoint. In a telling gesture, Dr. Coyne pointed his readers to Matzke's review as a definitive response to *Darwin's Doubt*. Currently a post-doctoral fellow at the National Institute for Mathematical and Biological Synthesis, Matzke has won renown for his tireless campaign to rebuke skeptics of evolutionary theory, a campaign going back to his days with the National Center for Science Education, an advocacy group in Oakland, California.

By his own account, Matzke is also a dizzyingly fast reader and writer. It was on June 19, 2013, the day after *Darwin's Doubt* was released and first made available for purchase, that Matzke published a

9,400-word critical review at *Panda's Thumb*.[1] Reading a book of this size and composing a review of that length all in little more than twenty-four hours would have to be recognized by anyone as a remarkable achievement. Challenged on how it was even possible, unless the review had been largely pre-written before he saw a copy, Matzke in a later post explained how he fit in his work on the review with other responsibilities, at lunchtime, in "snippets of the afternoon," and then by pulling an all-nighter. I, for one, am content to grant him this prodigy.

But what of the content of Matzke's critique?

Matzke's main criticism of *Darwin's Doubt* is that it failed to inform readers about how evolutionary biologists have been able to establish the existence of ancestors of the Cambrian animals using a method of phylogenetic analysis known as cladistics. According to Matzke, cladistic analysis has established the existence of "transitional" and "intermediate" forms between the animals that first arose in the Cambrian. In his view, cladistics has solved the problem of the missing ancestral fossils discussed in Part One (Chapters 1–7) of the book. As he asserted, "phylogenetic methods can establish, and have established, the existence of Cambrian intermediate forms, which are *collateral* ancestors of various prominent living phyla."[2] Matzke argued that my failure to inform readers of this disqualified the book from serious consideration as an analysis of the Cambrian explosion.

Of course, in making this argument Matzke scarcely addresses the central argument of my book: the problem of the origin of biological information. Neither does he offer any serious rebuttal to my argument in Chapter 11 of *Darwin's Doubt* showing that his 2004 article (co-authored with Alan Gishlick and Wesley Elsberry) failed to solve that problem.[3] As I showed in that chapter, Matzke and his colleagues at best described several mechanisms by which pre-existing genes, rich in *pre-existing* genetic information, can be shuffled and recombined.

In Chapter 11, and in the whole second part of my book (Chapters 8–14), I show that what most needs to be explained about the Cambrian

explosion is, essentially, a question of biological engineering—in particular, what *caused* the origin of the information necessary to specify the
novel animal structures and architectures that arose in the Cambrian.
Cladistics, by contrast, is a method of taxonomic classification, which,
like all such methods, takes these structures (or characters) as givens,
without considering how they were caused. Thus, cladistics bypasses the
problem of greatest interest.

Even so, Matzke did challenge a key secondary argument of the
book, namely, its claim that the absence of discernible ancestral forms
in the Precambrian fossil record represents a mystery from the neo-
Darwinian point of view. As I have noted, neo-Darwinism depicts the
history of life as a gradually unfolding branching tree in which all forms
of complex animal life arise by descent with modification from simpler
ancestral precursors. Now, this depiction of the history of life may be
true or false, but as an empirical claim, it cannot support itself. For that,
evidence is required. If the evidence is not forthcoming, however—if, for
instance, the fossils documenting the many morphological transformations required by this historical thesis are missing from the paleontological record—then simply restating (or presupposing) the thesis will
do nothing to repair that evidential defect. The question, therefore, is
exactly what *evidential* support does cladistics provide for the Darwinian
picture of the history of animal life—in particular, does it provide evidence for the existence of the presumed ancestors of the Cambrian animals that the fossil record does not document? As noted, Matzke claims
that cladistic analysis can establish, and has established, the existence of
various kinds of ancestors of the Cambrian animals.

But is this so?

Some Background

DARWIN'S DOUBT makes its case for the reality of the Cambrian explosion chiefly, but not entirely, on the basis of the fossil record. Representatives of twenty-three of the roughly twenty-seven fossilized animal
phyla (and of the roughly thirty-six total animal phyla) are present in
the Cambrian fossil record. Twenty of these twenty-three major groups

of animals make their first appearance in the Cambrian period with no discernible ancestral forms present in either earlier Cambrian or Precambrian strata. For the vast majority of the Cambrian animals, the evidence from paleontology suggests geologically abrupt appearance—an explosion (see Chapters 2–4 of *Darwin's Doubt*).[4]

In his review, Matzke insisted that *other* evidence nevertheless establishes the existence of the Cambrian intermediates or transitional forms. To make this claim he does not rely on any of the most common arguments against the reality of the Cambrian explosion. He does not claim that the Ediacaran organisms represent plausible ancestral forms of the Cambrian animals (see Chapter 4 of *Darwin's Doubt*); nor does he claim that these ancestral forms were not preserved because they were too small or too soft (see Chapter 3 of *Darwin's Doubt*); nor does he rely on phylogenetic reconstructions based on comparative gene sequences to establish Precambrian ancestors as advocates of deep divergence, for example, have done (see Chapters 5 and 6 of *Darwin's Doubt*). All of these proposals my book addresses and refutes.

Instead, Matzke invokes a more recently developed but arguably even less plausible approach to explaining away the absence of presumed ancestral forms. Matzke argues that phylogenetic reconstructions based on cladistic analysis establish the presence of intermediates and transitional forms that do not appear in the fossil record. *Darwin's Doubt* critiques this proposal only in passing (see page 60 of *Darwin's Doubt*), and instead provides an extensive critique of more commonly used methods of reconstructing evolutionary history based upon comparative analyses of DNA sequences. But my book did not devote the space to cladistics that Matzke thought it deserved. As Matzke argues:

> Meyer never presents for his readers the point that cladistic analyses reveal the order in which the characters found in living groups were acquired, nor the fact that stem taxa are the transitional fossils the creationists are allegedly looking for. And he especially avoids giving his readers any real sense of the number of transitional forms we know about for some groups, and the detail known about their relationships and about the order in which the characters of modern groups originated.[5]

Matzke also claims that *Darwin's Doubt* makes two significant errors regarding the classification of Cambrian animals. He claims that the book incorrectly refers to *Anomalocaris* as an arthropod, whereas, he argues, they are actually "stem-group" arthropods. He also claimed that the book incorrectly referred to Lobopodia as a phylum since, in his opinion, it represents a paraphyletic group (a group which contains some, though not all, descendants of the common ancestor of a group), likely encompassing the extant phyla Tardigrada and Onychophora. Matzke insisted that these alleged "basic errors" demonstrated my "ignorance" of systematics.

A few days later, my colleague Casey Luskin replied to Matzke (see Chapter 3 of the present book, "Rush to Judgment"), pointing out that *Darwin's Doubt* actually included two chapters with lengthy critiques of attempts to reconstruct phylogenetic histories using the similar technique of comparative sequence analysis, as well as a discussion of the distinction in cladistics between stem and crown groups. Indeed, in *Darwin's Doubt* I explain why making the distinction between stem and crown groups does not help explain what caused the Cambrian explosion or the origin of the biological information and anatomical characters that arose in it. Luskin also noted that many Cambrian scientific authorities have called *Anomalocaris* (and other members of its family, the anomalocaridids) "arthropods" of one type or another,[6] while other top authorities—including J. Y. Chen, James Valentine, and Douglas Erwin—have designated Lobopodia as a phylum.[7] Moreover, he noted, that in my book I acknowledge the uncertainty about the classification of anomalocaridids by describing them as "either arthropods or creatures closely related to them" (see *Darwin's Doubt*, page 53). Indeed, the very paper Matzke cited in recounting the history of phylogenetic analysis of Cambrian fossils states, "*Anomalocaris* is now recognized as an arthropod."[8] *Darwin's Doubt* did not discuss whether anomalocaridids were true arthropods or just stem group arthropods, but, as Luskin pointed out, the book does correctly note that they are generally regarded as arthropods of some type.

This set Matzke off again. That's just the point, he argued, in another lengthy response.[9] The difference between stem and crown groups is, he asserted, crucially important to reconstructing evolutionary histories. According to Matzke, that *Darwin's Doubt* didn't provide a detailed discussion of this distinction showed, again, that I didn't understand how evolutionary biologists do phylogenetic reconstructions using cladistic analysis.

So what is this debate all about? What exactly is cladistics? What are stem groups and crown groups, and does the distinction between the two allow evolutionary biologists to establish the existence of arthropod ancestors? And can cladistics establish the existence of the intermediates between, and the ancestors of, the Cambrian animals?

A Short Primer on Cladistics

CLADISTICS GENERATES branching patterns of relationships based upon an analysis of the number of "characters" (i.e., features, structures, or traits) shared by different types of organisms. The basic concept is simple. Systematists (experts in classification) examine a species to determine what characters it possesses. They then "score" whether the same characters are present in other presumably related taxonomic groups. After doing this for multiple characters and multiple species, they compare the number of characters that each species shares with other species. Species that share more characters are deemed to be more closely related than those that share fewer characters.

For cladists, not every shared character is important in their analysis. Cladistics is based upon comparing "shared derived" characters— those characters exclusively shared by all organisms in a group that can be traced (by inference) to the common ancestor of that group. Such characters are called synapomorphies. According to cladistics, the more shared derived characters that two species share, the closer their evolutionary relationship.

For example, let's assume that in a group of organisms there are five different characters of interest—A, B, C, D, and E. Let's also assume a simple distribution of characters where one organism possesses only character A, another has AB, another ABC, and so on. The resulting representation of the relationships between these organisms, called a cladogram, would look like this:[10]

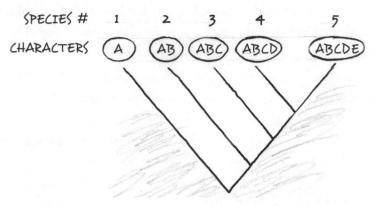

Figure 4-1. A simple cladogram with five species, showing how five separate characters, A, B, C, D, and E, are distributed among the species.

By using such a diagram and interpreting it as a representation of evolutionary history, evolutionary biologists can represent where various derived characters might have arisen, as seen in the tick marks on Figure 4-2 below:

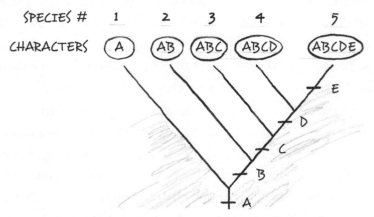

Figure 4-2. A simple cladogram with five species, showing how five separate characters, A, B, C, D, and E, are distributed among those species, as well as where they would have arisen on different lines of descent during the history of those groups.

Of course, reconstructing cladograms is almost never as simple as my idealized diagrams suggest. For any cladistic analysis, there will likely be many more characters than just the few in Figures 4-1 and 4-2. Systematists are often confronted with many characters within a group of species—the presence or absence of different anatomical structures, molecules, patterns of development, behaviors, and so forth—any of which, or any combination of which, could form the basis for producing cladograms. Thus, systematists face uncertainty about which characters (or combinations of characters) to include in their analyses, and further uncertainty about how to weight the characters they do include—at least, for those who practice "character weighting." (See discussion in this endnote.[11]) Once they have chosen those they regard as the most relevant (or "phylogenetically informative") shared derived characters, systematists feed the data about which animals possess which characters into an algorithm that generates the tree-like cladograms. These algorithms perform searches for *the* tree (or a set of trees)—among a huge number of possible trees—that provides the best overall fit with the data and involves the fewest number of separate evolutionary events (i.e., the fewest instances of gain or loss of characters).

Yet, as systematists include more characters in their analysis, the potential increases for generating inconsistent pictures of the history of life. So too does the need to apply subjective, *post hoc*, or theory-laden judgments about which characters to include, or about how to weight the different characters—at least, that is, if the algorithms are to produce reasonably coherent trees that conform to theoretical expectations about the nature of evolutionary change. An analysis of a group of species based upon one small set of characters may produce a clear, unambiguous cladogram. An analysis of the same group emphasizing a different set of characters can render an equally unambiguous branching tree pattern that is inconsistent with the first tree. An analysis including all the characters present in both data sets, however, can generate a complicated picture of evolutionary history in which some characters emerge or disappear on different branches independently. These patterns of character distribution are typically attributed to convergent evolution or loss of characters. (Alternately, the algorithm may identify many conflicting phylogenetic trees that are equally parsimonious.)

For example, imagine that in addition to characters A, B, C, D, and E in the figures above, a systematist also analyzes characters F and G. Imagine further that when characters F and G are included in the analysis, F occurs in species 1, 3, and 5, (but not on 2 and 4), and G appears in species 2, 4, and 5 (but not on 1 and 3), as seen in Figure 4-3. Explaining this pattern requires invoking multiple separate origins of the same characters (convergent evolution) and/or instances of character loss.

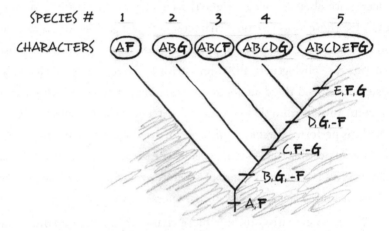

Figure 4-3. A cladogram with five organisms produced by a larger dataset
of characters, A, B, C, D, E as well as F and G. The characters shown
in bold represent those that must have arisen by convergent evolution
or those that were lost at some point. Minus signs in front of letters
indicate evolutionary events in which the characters were lost.

Since cladistics presupposes universal common descent, and since evolutionary biologists generally think the likelihood is low of characters arising multiple times on separate lines of descent, this type of analysis strives to minimize the number of such unexpected evolutionary events (especially separate origins of the same characters) necessary to explain the observed distribution of characters. This attempt to generate a tree, requiring the least number of steps, is called maximizing parsimony. However, maximizing parsimony (and minimizing the number of convergent events or loss) is frequently difficult as systematists include more characters in their analysis.

As noted, cladograms are constructed to take into account only shared derived characters. Groups that include species that have all of the shared derived characteristics that define a certain group (such as arthropods, for example) are called crown groups. Organisms that have some, though not all, of the shared derived characteristics defining the crown group are said to belong to the stem group. In Figure 4-2, for example, if traits A, B, C, D, and E are the shared derived characteristics which define the phylum ABCDE, then species with those characteristics are part of that crown group. However, other species possessing characters AB, ABC, and ABCD would be said to be members of the stem group of ABCDE.

Matzke thinks cladistic methods can establish evolutionary history, including both the sequence in which the characters defining a crown group arose, and the existence of various intermediates of the Cambrian animals. He claims that if paleontologists find an animal that shares some, but not all, of the shared characters that define a crown-group, then that animal can provide evidence for the existence of some intermediate—what he calls "a collateral ancestor"—of the crown group. Matzke, therefore, thinks that by distinguishing stem and crown groups, evolutionary biologists can establish intermediates—including intermediates between the Cambrian animals. This conviction explains why he reacted so negatively to Luskin's observation that *Darwin's Doubt* hadn't engaged the debate about whether the anomalocaridids represent stem group or crown group arthropods.

Instead, in his view, "the arthropods are *instructive*" in how cladistics can establish the existence of intermediates between the Cambrian forms, since "when fossils are analyzed cladistically, we typically discover a bunch of species that morphological characters place below the crown—i.e., 'stem groups.'" He thinks making this distinction helps evolutionary biologists, "learn the basics of how 'body plans' originated by using cladistics (or more sophisticated methods) to estimate the order and timing of each character change found in the crown group." He claims that "cladistic analyses reveal the order in which the characters found in living groups were acquired." Specifically, he argues that because *Anomalocaris* possesses some, though not all, of the features of true arthropods, "it is one of many fossils with transitional morphology *between* the crown-group arthropod phylum, and the next closest living crown group, Onychophora (velvet worms)."

Matzke's claims notwithstanding, there are several reasons to doubt that cladistic methods, and the distinction between stem and crown groups, can establish *ancestral* precursors or ancestral intermediates to the Cambrian animal groups, including putative ancestors of the arthropods. (More on whether Matzke actually claims that below.)

Ghost Lineages and Chronological Inversions

MATZKE THINKS that cladistic analysis of *Anomalocaris* and other fossils reveals some kind of "intermediate," "transitional," or "ancestral" arthropods, effectively solving the mystery of the missing ancestral forms of these animals. Yet using cladistics to infer such ancestral arthropods requires postulating "ghost lineages" that imply the existence of still *more* missing fossils. The need to invoke hypothetical ghost lineages commonly arises when evolutionary biologists attempt to use cladistics to infer ancestors otherwise unattested by the fossil record. The reason for this is that the fossil record often reveals so-called stem groups arising contemporaneously with, or even after, crown groups. Theropod dinosaurs provide a classic example of this problem. They first appear in the fossil record millions of years after the birds that allegedly evolved from them.[12] Similarly, many supposed members of stem group arthropods appear in the Cambrian fossil record *contemporaneously* with, or *after*, members of the crown group arthropods that they supposedly preceded.

The anomalocaridids (and other species) that, according to Matzke, represent Cambrian intermediates illustrate this kind of chronological problem. Recall that it was my supposed misclassification of these animals to which Matzke objected in the first place. In response, he provided a cladogram from a 2012 paper lead-authored by David Legg,[13] then at the Natural History Museum of London, showing *Anomalocaris* as a stem group arthropod, which Matzke would classify as intermediate or transitional to true arthropods. Yet, arthropod specialist Gregory Edgecombe reports that Radiodonta, the larger group to which *Anomalocaris* belongs, appears in the fossil record at the same time as true arthropods, *not before*.[14] Indeed, as seen in Figure 4-4, reproduced from a 2010 paper by Edgecombe, *none* of the stem group arthropods appear in the fossil record *before* the appearance of their supposed evolutionary descendants, the crown group (or "true") arthropods. As a 2013 article by Edgecombe and Legg explains, *Anomalocaris* appears at about the same time as true arthropods—*not before*—"in both the Burgess Shale in Cambrian Stage 5 in Canada (on the palaeocontinent Laurentia) and in the Chengjiang biota in Cambrian Stage 3 in China."[15]

Similarly, the 2012 paper by Legg and colleagues that Matzke cited reported an analysis of the stem group arthropod *Nereocaris*. Matzke posts a cladogram from the paper showing this animal as an intermediate between *Anomalocaris* and true arthropods with a caption purporting to show "the phylogenetic position of *Nereocaris*."[16] But such a claim implies a chronological inversion, since *Nereocaris* is known from the Tulip Beds locality of the Burgess Shale, dated at about 505 million years ago—some 15 million years *after* the first true arthropods appeared.[17]

Why are such inversions a problem? For evolutionary biologists to produce phylogenetic trees depicting evolutionary history consistent with cladistic analysis in cases involving inversions, they must draw long branches representing lineages for which they lack fossil representatives. In the case of the arthropods, these ghost lineages must stretch well back in time to connect to the hypothetical ancestor of all the stem and crown group arthropods. Figure 4-4 from Edgecombe (2010) depicts this problem. In that figure, the thick black lines represent time periods from which fossils of various (both stem and crown group) arthropods are known, as well as other groups that are supposedly close relatives of arthropods. The thin black lines represent animals inferred based on cladistic analysis—animals that are *not* found in the fossil record. Note that all of the putative ancestors of arthropods that might link them to other groups are represented by thin black lines. Indeed, none of the supposed *ancestral* stem arthropods, or their evolutionary histories, or

related non-arthropod ancestral groups from which arthropods supposedly evolved, are documented in the fossil record.

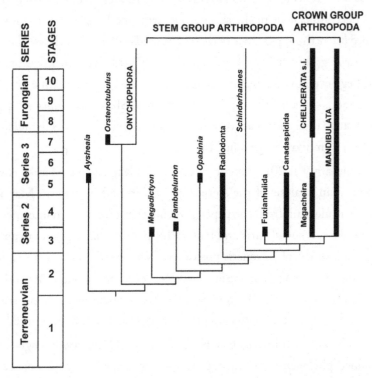

Figure 4-4. The fossil record of stem group and crown arthropods, and other
related animals, plotted against the Cambrian time scale. Thick black lines
represent known fossil record. Thin black lines represent presumed evolutionary
history for which there are no known fossils. Reprinted with permission of
Elsevier from Gregory D. Edgecombe, "Arthropod phylogeny: An overview from
the perspectives of morphology, molecular data and the fossil record," *Arthropod
Structure & Development*, 39 (2010): 74-87, Figure 1. © Elsevier, 2010.

For example lobopods (represented in Figure 4-4 by *Orstenotubulus*,
and possibly *Aysheaia*) are thought to be closely related to the ancestors
of arthropods, if not directly ancestral to arthropods, but as we see in
Figure 4-4, they don't appear until millions of years *after* the first true arthropods. Yet Matzke claims that "the arthropod and velvet-worm phyla
[Onychophora] evolved from lobopods, and lobopods contain a whole
series of transitional forms showing the basics of how this happened."[18]
Do we see this "whole series" in the fossil record? Richard Fortey com-

ments in *Science* that "Onychophora (velvet worms) were probably the most closely related group to the arthropods as a whole; this group and the arthropods must have diverged... in the Precambrian." Rather than finding a "whole series of transitional forms," however, Fortey calls this "earlier" evolution of arthropods "Precambrian hidden history," acknowledging that "fossils of these alleged ancestral arthropods are lacking."[19] To put some numbers on the problem, a 2011 paper in *Science* used molecular clocks to date the split of arthropods and Onychophora to over 600 million years ago,[20] but neither group appears in the fossil record until around 521 million years ago or later. That's at least 80 million years of "hidden history" of arthropods, with a group (the lobopods) representing supposed arthropod "collateral ancestors" appearing *after* arthropods.

The case of *Schinderhannes bartelsi*, an anomalocaridid known only from rocks of the lower Devonian, provides another striking example of such an inversion. Indeed, when touting the findings of cladistic analysis, Matzke might have easily cited a 2009 paper in *Science* reporting the discovery of *Schinderhannes*, which called *Schinderhannes* a "stem lineage" arthropod, and included cladistic analysis making it appear intermediate (by Matzke's standard) between *Anomalocaris* and true arthropods.[21] Or, he might have cited a cladogram from a 2011 paper in *Nature* showing *Schinderhannes* as one of the closest relatives (what Matzke might call a "collateral ancestor") to crown-group arthropods.[22] But had he done so, it would have again highlighted the need to formulate ghost lineages to generate a coherent phylogenetic tree. Indeed, *Schinderhannes* appeared *over 100 million years after* the first true arthropods are found. This chronological inversion requires the postulation of a ghost lineage of over 100 million years to place *Schinderhannes* in a correct phylogenetic relationship to other arthropods.

So does cladistic analysis—showing that *Anomalocaris* (and some of its close relatives) are lacking some characters of crown group arthropods—establish that *Anomalocaris* represents an *ancestral* intermediate between onychophorans and crown group arthropods? Matzke can make this argument only by assuming that these stem group arthropods (or their relatives) existed *before* crown group arthropods first appeared. But since the fossil record does not document the existence of *Anomalocaris* or its relatives in the earlier fossil record, those using cladistics to infer the evolutionary history of crown group arthropods must also posit ghost lineages of earlier fossil *Anomalocaris*-like ancestors reaching back into the record long before crown group arthropods appeared. Did such a sequence really exist? Who knows? But it hardly solves the problem of missing fossil ancestors of the Cambrian animals to use cladistics to posit a phylogenetic hypothesis that requires, as a condition of its plausibility, the postulation of ghost lineages representing still more missing fossils.

Wrong or Irrelevant

THERE IS another problem with Matzke's use of cladistics. Many cladists themselves do not think that cladograms necessarily indicate anything about evolutionary history. Instead, they regard them as tools for classifying different taxa. Matzke himself acknowledges at least one important limitation on what cladograms can reveal about evolutionary history by conceding that "phylogenetic methods as they exist now can only rigorously detect sister-group relationships, not direct ancestry." This weaker claim about what cladistics can tell us is much easier to defend given the paucity ancestral forms in the Precambrian fossil record. But if that is *all* that Matzke means to claim about what cladistics can establish, then the significance of his argument evaporates.[23]

Oddly, in his discussion of the use of cladistics in phylogenetic reconstruction, Matzke never defines exactly what he means by an "intermediate." Does he mean an intermediate only in the sense of an animal possessing some but not all of the features of the crown group? An anatomical intermediate? Or does he mean a true *ancestral* intermediate of the kind that *Darwin's Doubt* argues is missing? Matzke does not specify, though presumably he would insist that he does not necessarily mean to imply that the anomalocaridids were the direct *ancestral* precursors to the trilobites or other crown group arthropods. Thus, he uses the ambiguous word "collateral" as a modifier to the word ancestor when he states:

"I claimed that phylogenetic methods can establish... *collateral* ancestors of various prominent living phyla."[24]

It's not entirely clear what Matzke means by "collateral ancestor," because he never defines the term. Indeed a search of PubMed for the term "collateral ancestor" reveals virtually nothing in the technical literature: only three hits were returned for "collateral ancestor" or "collateral ancestors," and in none was the term used in the way that Matzke did. The term does, however, have meaning in a legal context. As the online *Encyclopedia of Genealogy* explains, "Collateral ancestor is a legal term referring to a person [who is] not in the direct line of ascent, but is of an ancestral family. This is generally taken to mean a brother or sister of an ancestor (hence a 'collateral ancestor' is never an ancestor of the subject)."[25] In other words, a collateral ancestor is not a direct, actual, or common ancestor, and thus does not solve the problem of the missing fossils of the common (or direct) ancestors of the arthropods or other major Cambrian groups.

Matzke might reply that what I've said misunderstands the subtlety of *his* position. He might say that he is not claiming that the anomalocaridids are the *ancestors*, or at least the *direct ancestors*, of arthropods, or that members of stem groups are necessarily the direct ancestors of specific members of crown groups. Instead, he might say that he only means to affirm that they are intermediates in the sense of possessing some but not all of the characters of the crown group, and that they are *collateral* ancestors, meaning that they reside somewhere on the evolutionary tree below the crown group and thus are in some way related to the direct ancestors of the crown groups.

But, again, if that is *all* that Matzke means, then surely his use of cladistics does not solve the problem highlighted in the first third of *Darwin's Doubt*. Recall that my first seven chapters argued that neither fossil nor genetic evidence establishes the existence of *ancestral* precursors for most of the Cambrian animals. If Matzke's intermediates are not direct *ancestral* precursors of the Cambrian animals, then they do not provide the missing intermediates highlighted by *Darwin's Doubt*. If, on the other hand, Matzke *is* claiming that cladistics resolves the mystery of missing ancestral fossils, then he is simply wrong, because the temporal order of the appearance of character states in the Cambrian requires the postulation of ghost lineages representing still other missing ancestral fossils. Since Matzke never clearly defines what he means by an intermediate, it is not possible to establish upon which horn of this dilemma his position ultimately founders. Either way, cladistics-based phylogenetic hypotheses do not solve the problem of missing ancestral fossils.

In any case, there are still further difficulties with his position.

Begging the Question

IN 2012, molecular biologist Michael Syvanen observed in *Annual Review of Genetics* that "one needs to be continually reminded that submitting multiple sequences (DNA, protein, or other character states) to phylogenetic analysis produces trees because that is the nature of the algorithms used."[26] The same can be said about analyses of shared derived characters and cladograms constructed on the basis of cladistic analysis (and other forms of character-based phylogenetic analysis). For those who regard cladograms as depicting real events in evolutionary history, the algorithms used during phylogenetic reconstructions and cladistic analysis presuppose, rather than demonstrate, the common ancestry of the groups they analyze. Indeed, the assumption of common ancestry is inherent in the method of cladogram construction—at least for those who regard such trees as representing evolutionary history, rather than mere classificatory devices. As University of Wisconsin philosopher of biology, Elliott Sober states, when evolutionary biologists construct cladograms, "the typical question is which tree is the best one, not whether there is a tree in the first place."[27]

Furthermore, if one interprets the results of cladistic analysis as an indicator of evolutionary history, then the number of shared derived characters represents a measure of the *historical* relatedness of two or more groups. Viewed this way, cladistic analysis assumes that more shared derived characters, indicate—all other things being equal—a closer evolutionary relationship and a more shallow divergence *from a common ancestor.* Conversely, it also assumes that fewer shared derived characters indicate a more distant evolutionary relationship and a deeper divergence *from a common ancestor.* Thus, interpreting the results of cladistic analysis historically entails the assumption that each of the groups analyzed evolved from a common ancestor. It presupposes, rather than demonstrates, the existence of such ancestors.

One sees this assumption of common descent in nearly every phase of cladistics and other similar forms of character-based phylogenetic analysis. When systematists choose which characters to include and which to exclude, they make judgments about which characters are most "phylogenetically informative."[28] In practice, this means selecting those characters that are most likely to generate congruent treelike patterns requiring the fewest evolutionary events. Judgments about how to weight different characters or about which species to place at the base of a given phylogenetic tree (how to "root" the tree) are made with similar considerations in mind, and are always informed by the background assumption of common ancestry.

In Chapters 5 and 6 of *Darwin's Doubt* I made the same point about the use of comparative sequence analysis to generate phylogenetic trees. By presupposing that degrees of difference indicate time elapsed since divergence from a common ancestor, those methods also presuppose, rather than demonstrate, a common ancestor. Those who cite either comparative sequence or character-based phylogenetic analyses to establish the existence of such ancestral forms, elide a simple point of logic. No method that presupposes the truth of a proposition can be used to prove or establish the truth of that same proposition without begging the question. By asserting that phylogenetic reconstructions based on cladistic analysis "can establish, and have established, the existence of Cambrian intermediate forms" on the animal tree of life, Matzke relies on precisely such a question-begging method.

Index of Inconsistency

THERE IS another reason not to regard cladograms as representations of evolutionary history as opposed to classificatory devices. Characters that appear homologous do not always reflect common ancestry. Even evolutionary biologists who assume universal common descent recognize this. As I noted in Chapter 6 of *Darwin's Doubt*, the assumption that shared characteristics result from common ancestry commonly breaks down in the case of those characteristics thought to have arisen via "convergent" evolution. The endnotes in my book cite textbooks on phylogenetic methods that acknowledge the *assumption* of common ancestry, as well as the difficulty that assumption can pose for reconstructing phylogenetic trees using cladistic analysis. One source states:

> Cladistics can run into difficulties in its application because not all character states are necessarily homologous. Certain resemblances are convergent—that is, the result of independent evolution. We cannot always detect these convergences immediately, and their presence may contradict other similarities, 'true homologies' yet to be recognized. Thus, we are obliged to assume at first that, for each character, similar states are homologous, despite knowing that there may be convergence among them.[29]

The problem of convergent evolution is rampant in cladistic studies of Cambrian arthropods. To see why, consider how often it is necessary to invoke convergent evolution (or loss) to explain the distribution of characters in many cladograms. Evolutionary biologists have quantitative ways of measuring this. One method involves calculating the consistency index (CI). It is a statistical measure of how often the assumption of common ancestry succeeds (or fails) in explaining the distribution of shared biological characters. It is calculated simply by taking the minimum number of evolutionary events required by the overall dataset (which is equivalent to the total number of characters being studied) and dividing by the number of evolutionary events implied by a given tree. A high CI (closer to 1) indicates the characters are naturally distributed in a treelike pattern without having to invoke additional evolutionary events. A lower CI (closer to 0) means that it is difficult to explain the distribution of characters without invoking many instances of convergent evolution (or loss).

In his review of my book, Matzke insisted on assessing the "statistical support" for an evolutionary tree as a prerequisite to challenging its veracity. Let's therefore consider the consistency indices of the two cladograms that he cites, each of which, according to Matzke, establishes definitive evolutionary relationships among Cambrian arthropods. One cladogram from Legg and his colleagues (2012) has a CI of 0.565,[30] meaning that about 43.5 percent of the time, a given character was *not* distributed in a tree-like pattern. To put it another way, 43.5 percent or so of the time, the assumption of common ancestry failed to explain the distribution of a character. Even worse, the other cladogram Matzke cited has a CI is 0.384, which the original authors admit was "rather low."[31] This is striking: roughly 61.6 percent of the time, the shared characters in these groups did *not* result from descent with modification from a common ancestor. That is, in 61.6 percent of the cases, the assumption of homology failed to explain the distribution of characters.

If an assumption fails more often than it holds true, is it justified? Whatever the answer, it's clear that the characters of the Cambrian arthropods often fail to fit the tree-like pattern required by universal common descent. This further undermines Matzke's claim that cladistics establishes the existence of the ancestors implied by trees generated from this method.

Conflicting Trees

THERE IS another problem with treating cladograms as depictions of evolutionary history. The same set of characters will often generate many equally parsimonious cladograms. Depending upon the choices that one makes about how to do a character-based phylogenetic analysis—which characters to emphasize as homologous, how strongly to weight characters, which computer programs to use in the analysis, whether to take into account paleontological data about specific taxa, how to "root"[32] trees, and so on—one can generate many different, conflicting trees from the same dataset. Officially a "pure cladist" will not weight characters in his analysis (see endnote 11). Nevertheless, in practice, many evolutionary biologists doing character-based phylogenetic analysis, including many who consider themselves cladists, do weight characters differently—causing phylogenetic algorithms to generate different, and sometimes conflicting trees in response to different choices about *how* to weight characters.

For example, some evolutionary biologists think that both lobopods and radiodonts (the order that includes anomalocaridids) are closely related to arthropods. Lobopods can have arthropod-like legs, but lack arthropod-like heads and eyes. Radiodonts lack legs, but can have arthropod-like heads and eyes. Depending upon which characters are weighted more heavily, phylogenetic analysis will generate starkly different and incompatible trees, showing, in one case, lobopods arising first and radiodonts later, and in another case, just the reverse.[33] Decisions about other factors—how to root trees, whether to take paleontological data into account, and so on—can also result in multiple, conflicting trees.

Evolutionary biologists who interpret these differing trees as depictions of evolutionary history immediately face a difficult problem. Which of the conflicting treelike diagrams reflects the true evolutionary history of a given group? If the same raw data generate many conflicting trees, how can we say that cladistic data are sending a clear historical signal?

Matzke ignores this problem by affirming a single unequivocal history of the arthropods. He confidently asserts that "the arthropod and velvet-worm phyla *evolved from lobopods* and lobopods contain a whole series of transitional forms" in which *Anomalocaris* has "transitional morphology between the crown-group arthropod phylum and the next closest living crown group, Onychophora (velvet worms)."[34] Nevertheless, the history of the arthropods that Matzke affirms represents just one of many possible histories allowed by character-based phylogenetic analysis. In affirming a particular historical progression as *the* correct history, Matzke represents phylogenetic analysis—and indeed, cladistics, narrowly construed—as far more capable of establishing a definitive picture of the history of animals than the relevant scientific literature indicates.[35] Thus, for example, Douglas Erwin and James Valentine's 2013 book *The Cambrian Explosion* calls arthropod origins "far from settled" and even "problematic."[36] Or as Edgecombe writes, "Arthropod phylogeny is sometimes presented as an almost hopeless puzzle wherein all possible competing hypotheses have support."[37] Even Legg and his colleagues, whose articles Matzke cites in support of his critique of *Darwin's Doubt*, note: "The origin of arthropods is a contentious issue... [T]here is little consensus regarding the details of their origins."[38]

Conflicting Histories

Furthermore, even in an idealized case where cladistics generates one cladogram that is clearly the most parsimonious, that tree itself necessarily corresponds to many separate possible evolutionary histories. Oddly, one scholar Matzke cites in his critique of *Darwin's Doubt* makes exactly this point. To support his claim that cladistic analysis enables evolutionary biologists to establish the "evolutionary history" of animal groups and the existence of "transitional forms," Matzke cites as authoritative a 2008 paper by historian of paleontology Keynyn Brysse. In Figure 3 of her article (reproduced here as Figure 4–5), Brysse shows graphically why cladistic analysis does *not* establish definitive evolutionary history. That figure demonstrates how one simple cladogram representing just three character states is equally consistent with *six* different evolutionary histories. In the caption to the figure, she explains that "the cladogram is not an expression of ancestor-descendant relationships; it is not a phylogenetic tree."[39] In the article, she further explains the limitation of cladistics:

> Cladograms depict sister groups—taxa that are thought to be each other's closest relative—but do not show ancestors and descendants. It is possible to use a cladogram as the basis for constructing an evolutionary tree, but any given cladogram is *usually consistent with multiple phylogenetic trees.* This means that according to cladistics at least, the correct ancestral–descendant relationships among fossils are always underdetermined by the available evidence. In other words, the very goal of evolutionary systematics—the determination of ancestor–descendant rela-

tionships—is on the cladistic method not just unattained but unattainable.[40]

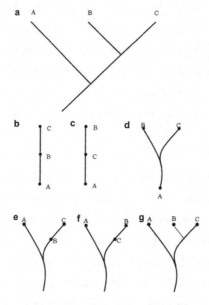

Figure 4-5. Brysse's Figure 3, showing that for every cladogram, there are many possible corresponding evolutionary histories. As Brysse explains: "The cladogram at the top (a) shows that among three related taxa, A, B, and C, B and C are more closely related to each other than either is to A. There is no evidence, however, to distinguish the possible ancestor-descendant relationships among the three taxa. For example, A might have evolved into B, and then into C (b); A might have evolved into C and then into B (c); B and C might both be descendants of A (d); A and B might be descendants of a common ancestor, with B evolving into C (e); A and C might be descendants of a common ancestor, with C evolving into B; or finally, B and C might have evolved from a common ancestor which itself shares a common ancestor with A. As these examples illustrate, the cladogram is not an expression of ancestor-descendant relationships; it is not a phylogenetic tree. The fact that at least six trees can be drawn from a single cladogram is evidence that the correct tree is underdetermined by the cladistic data. As cladists argue, then, there is only enough reliable information available to construct cladograms, not trees." Reprinted with permission of Elsevier from Keynyn Brysse, "From weird wonders to stem lineages: the second reclassification of the Burgess Shale fauna," *Studies in History and Philosophy of Science Part C: Studies in History and Philosophy of Biological and Biomedical Sciences*, 39 (2008): 298–313, Figure 3. © Elsevier, 2008.

Of course, Matzke does acknowledge, in some places at least, that cladistics can only establish sister groups and not direct ancestors. But throughout his review, he also claims that cladistics can establish the sequence in which characters arose, the branching order of different animal groups on the tree of life, and the existence of "transitional forms" leading to the Cambrian animals—in short, many key aspects of "evolutionary history." Yet, a critical source that he cites shows that establishing definitive historical claims about specific animal groups using cladistics is, to say the least, problematic.

Cladistics Cannot Determine Causes

BRYSSE'S PAPER is instructive in another respect. She argues that cladistics cannot establish anything about the *processes* that might have produced the characters represented, and the patterns depicted, in cladograms. After recounting the history of classification of Cambrian organisms, she argues that cladistics does nothing to solve the mystery of the Cambrian explosion. Brysse explains that cladistics allowed evolutionary scientists "to construct clearly stated hypotheses about the relationships among the organisms under examination" but not "to investigate the tempo and mode of evolution." Instead, cladistic analysis necessarily "ignores"[41] questions about the causal processes that generate evolutionary novelty.[42] To emphasize her point Brysse cites another authority, Henry Gee, who puts the point succinctly: "Cladistics is concerned with the pattern produced by the evolutionary process; it is not concerned with the process that created the pattern, or the swiftness or slowness with which that process acted."[43]

Cladistics describes patterns of relationships among organisms; it provides tools for classifying organisms. It *might* also suggest historical reconstructions of evolutionary history *if* its question-begging assumptions in that context are granted. But it cannot determine what caused the patterns of relationship depicted by cladograms or what *caused the origin of the complex animal features* that it analyzes. For this reason, cladistics cannot be used to rebut the central argument of my book, which addresses precisely the question of what *caused* the Cambrian animals to arise.

And that is why Matzke's review of *Darwin's Doubt* fails to address the central argument of my book. Cladistics does not, and cannot, offer any explanation of what caused the Cambrian animals to come into existence. Nor can it account for the origin of the genetic and epigenetic information necessary to produce them.

Notes

1. Nick Matzke, "Meyer's Hopeless Monster, Part II," *Panda's Thumb*, June 19, 2013, http://pandasthumb.org/archives/2013/06/meyers-hopeless-2.html. See also Nick Matzke "Luskin's Hopeless Monster," *Panda's Thumb*, June 27, 2013, http://pandasthumb.org/archives/2013/06/luskins-hopeles.html.

2. Matzke, "Luskin's Hopeless Monster." See also Matzke, "Meyer's Hopeless Monster, Part II."

3. See also Casey Luskin, "Does Natural Selection Leave 'Detectable Statistical Evidence in the Genome'?," *Evolution News & Views*, August 7, 2013, http://www.evolutionnews.org/2013/08/does_natural_se075171.html.

4. See Casey Luskin's response to Nick Matzke responding reinforcing this same claim. Casey Luskin, "How 'Sudden' Was the Cambrian Explosion? Nick Matzke Misreads Stephen Meyer and the Paleontological Literature; New Yorker Recycles Misrepresentation," *Evolution News & Views*, July 16, 2013, http://www.evolutionnews.org/2013/07/how_sudden_was_074511.html.

5. Matzke, "Meyer's Hopeless Monster, Part II."

6. The following sources identify *Anomalocaris* as a type of arthropod: John R. Paterson et al., "Acute vision in the giant Cambrian predator *Anomalocaris* and the origin of compound eyes," *Nature* 480, no. 7376 (December 8, 2011), http://www.nature.com/nature/journal/v480/n7376/full/nature10689.html; Ed Yong, "The sharp eyes of *Anomalocaris*, a top predator that lived half a billion years ago," *Discover*, December 7, 2011, http://blogs.discovermagazine.com/notrocketscience/2011/12/07/anomalocaris-sharp-eyes-predator/; Benjamin M. Waggoner, "Phylogenetic Hypotheses of the Relationships of Arthropods to Precambrian and Cambrian Problematic Fossil Taxa," *Systematic Biology* 45, no. 2, (1996): 190–222, http://sysbio.oxfordjournals.org/content/45/2/190.abstract; Jianni Liu et al, "A large xenusiid lobopod with complex appendages from the Lower Cambrian Chengjiang Lagerstätte," *Acta Palaeontologica Polonica* 51, no. 2 (2006): 215–222, http://www.app.pan.pl/article/item/app51-215.html; Charles R. Marshall and James W. Valentine, "The importance of preadapted genomes in the origin of the animal bodyplans and the Cambrian explosion," *Evolution* 64, no. 5 (May 2010): 1189–0201, http://www.ncbi.nlm.nih.gov/pubmed/19930449; Graham E. Budd and Søren Jensen, "A critical reappraisal of the fossil record of the bilaterian phyla," *Biological Reviews of the Cambrian Philosophical Society* 75, no 2 (May 2000): 253–95, http://www.ncbi.nlm.nih.gov/pubmed/10881389; Xianguang Hou et al., *The Cambrian Fossils of Chengjiang, China: The Flowering of Early Animal Life* (Malden, MA: Blackwell, 2004), 94.

7. The following sources identify Lobopodia as a phylum: Douglas Erwin *et al.*, "The Cambrian Conundrum: Early Divergence and Later Ecological Success in the Early History of Animals," *Science* 334, no 6059 (November 25, 2011): 1091–1097, http://www.sciencemag.org/content/334/6059/1091.full.html; T. Cavalier-Smith, "A revised six-kingdom system of life," *Biological Reviews of the Cambridge Philosophical Society* 73, no 3 (August 1998): 203-66, http://onlinelibrary.wiley.com/doi/10.1111/j.1469-

185X.1998. tb00030.x/abstract; M. Alan Kazlev, "Panarthropoda: Lobopodia," *Palaeos* (July 15, 2002), http://palaeos.com/metazoa/ecdysozoa/panarthropoda/lobopodia.html (stating "The Lobopodia or Lobopoda are an evolutionary grade or phylum"); Xianguang Hou et al., *The Cambrian Fossils of Chengjiang, China*, 82.

8. Keynyn Brysse, "From weird wonders to stem lineages: the second reclassification of the Burgess Shale fauna," *Studies in History and Philosophy of Science Part C: Studies in History and Philosophy of Biological and Biomedical Sciences* 39, no. 3 (September 2008): 298–313.

9. See Matzke, "Luskin's Hopeless Monster."

10. For a classic and real example of such cladistic analysis see Niles Eldredge and Joel Cracraft, *Phylogenetic Patterns and the Evolutionary Process* (New York: Columbia University Press, 1980), 28. See also Vicki A. Funk and Quentin D. Wheeler, "Symposium: Character Weighting, Cladistics and Classification," *Systematic Zoology* 35, no. 1 (1986): 100–101.

11. "Weighting" refers to the practice of determining evolutionary relationships based on the characters that systematists decide are most important for defining a particular group. Many systematists do not regard character weighting as a part of cladistics, at least strictly speaking. See Brysse, "From weird wonders to stem lineages," 305. One distinctive of cladistics as a method of evolutionary systematics has been its attempt to eliminate subjectivity from its analytical methods. Accordingly, strict cladists reject differential weighting of characters as a means of producing the most parsimonious cladograms. Nevertheless, many evolutionary systematists who do character-based phylogenetic analysis do choose to weight some characters more heavily than others. Whether such weighting represents part of cladistics specifically, or just part of evolutionary systematics more generally, is partly a semantic issue and one that is controversial among practitioners. In any case, character-based phylogenetic analysis often does involve weighting characters and decisions about how to weight characters affect which of the many possible trees are generated by the phylogenetic algorithms.

12. See Robert A. Martin, *Missing Links: Evolutionary Concepts & Transitions Through Time* (Boston: Jones and Bartlett, 2004), 153 (stating theropods "all occur in the fossil record after Archaeopteryx and so cannot be directly ancestral"); Carl C. Swisher III, Yuan-qing Wang, Xiao-lin Wang, Xing Xu, and Yuan Wang. "Cretaceous Age for the Feathered Dinosaurs of Liaoning, China," *Nature* 400 (1999): 58–61, http://www.nature.com/nature/journal/v400/n6739/abs/400058a0.html.

13. David A. Legg, Mark D. Sutton, Gregory D. Edgecombe, and Jean-Bernard Caron. "Cambrian Bivalved Arthropod Reveals Origin of Arthrodization," *Proceedings of the Royal Society B* 279 (2012): 4699–4704, http://rspb.royalsocietypublishing.org/content/early/2012/10/03/rspb.2012.1958.

14. Gregory D. Edgecombe, "Arthropod Phylogeny: An Overview from the Perspective of Morphology, Molecular Data, and the Fossil Record," *Arthropod Structure & Development* 39 (2010): 74–87, http://www.sciencedirect.com/science/article/pii/S1467803909000541.

15. Gregory D. Edgecombe and David A. Legg, "The Arthropod Fossil Record," in *Arthropod Biology and Evolution: Molecules, Development, Morphology*, edited by A. Minelli et al., 393–415 (Berlin: Springer-Verlag, 2013), 394.

16. Legg et al., "Cambrian bivalved arthropod reveals origin of arthrodization," 4702.

17. Ibid.

18. Matzke, "Meyer's Hopeless Monster, Part II."

19. Richard Fortey, "Evolution: The Cambrian Explosion Exploded?" *Science* 293 (2001): 438–39, http://www.sciencemag.org/content/293/5529/438.

20. Erwin et al., "The Cambrian Conundrum," 1094.

21. Gabriele Kühl, Derek E. G. Briggs, and Jes Rust, "A Great-Appendage Arthropod with a Radial Mouth from the Lower Devonian Hunsrück Slate, Germany," *Science* 323 (February 6, 2009): 771–73, 771, https://www.sciencemag.org/content/323/5915/771. full.pdf.

22. See Figure 3, Paterson et al., "Acute vision in the giant Cambrian predator *Anomalocaris* and the origin of compound eyes," 239.

23. Arguably, organisms like *Anomalocaris* also make poor anatomical intermediates to arthropods as well, since they bear unique features—entirely ignored by cladistics— like strange flexible side-lobes, or a mouth with ringed teeth, which are foreign to crown-group arthropods. See Harry B. Whittington and Derek E. G. Briggs, "The Largest Cambrian Animal, Anomalocaris, Burgess Shale, British Columbia," *Philosophical Transactions of the Royal Society B* 309 (1985): 569–609, 604, http://rstb.royalsocietypublishing.org/content/309/1141/569 ("*Anomalocaris*... is unlike any known arthropod, particularly in the nature of the jaw apparatus and the close-spaced, strongly overlapping lateral lobes"); Brysse, "From weird wonders to stem lineages," 307 ("Cladistics, by contrast, gives no weight to unique characters"); Derek Briggs and Richard Fortey, "The Early Radiation and Relationships of the Major Arthropod Groups," *Science* 246 (1989): 241–43, 243, http://www.sciencemag.org/content/246/4927/241 ("The cladistic approach, on the other hand, focuses on shared characters. Unique attributes, autapomorphies, are of no use in assessing relationship and are consequently accorded little significance.").

24. Matzke, "Meyer's Hopeless Monster, Part II" (emphasis added).

25. Robert Shaw, "Collateral Ancestor," *Encyclopedia of Genealogy*, http://www.eogen. com/collateralancestor.

26. Michael Syvanen, "Evolutionary Implications of Horizontal Gene Transfer," *Annual Review of Genetics* 46 (2012): 339–56, 341, http://www.annualreviews.org/doi/ abs/10.1146/annurev-genet-110711-155529.

27. Elliott Sober and Michael Steele, "Testing the Hypothesis of Common Ancestry," *Journal of Theoretical Biology* 218 (2002): 395–408, 395, http://www.sciencedirect. com/science/article/pii/S0022519302930869.

28. This telltale phrase is remarkably widespread in the cladistics literature. See, for instance, Gregory D. Edgecombe, Gonzalo Giribet, Casey W. Dunn, Andreas Hejnol, et al., "Higher-level metazoan relationships: recent progress and remaining questions," *Organisms, Diversity, & Evolution* 11 (2011): 151–172, 152: "the complexity and number of potential phylogenetically informative characters often forced scientists to base their hypotheses on a few selected characters—e.g. larval ciliary bands, excretory systems, or embryology in the case of deep metazoan relationships—leaving out much other important information and carrying the risk of producing biased results based on homoplasies." See also Christopher W. Wheat and Niklas Wahlberg, "Phylogenomic Insights into the Cambrian Explosion, the Colonization of Land and the Evolution of Flight in Arthropoda," *Systematic Biology* 62, no. 1 (2013): 93–109.

29. Guillaume Lecointre and Hervé Le Guyader, *The Tree of Life: A Phylogenetic Classification* (Cambridge, MA: Harvard University Press, 2006), 16.

30. Legg et al., "Cambrian bivalved arthropod reveals origin of arthrodization."

31. Briggs and Fortey, "The Early Radiation and Relationships of the Major Arthropod Groups."

32. Rooting refers to the practice by which systematists decide which taxon among a group of taxa under analysis is most likely to have been the last common ancestor of the group and force a phylogenetic algorithm to generate only trees that reflect that decision.

33. See Jianni Liu, Michael Steiner, Jason A. Dunlop, Helmut Keupp, et al., "An armoured Cambrian lobopodian from China with arthropod-like appendages," *Nature* 470 (February 24, 2011), 526–530; Ross C. P. Mounce and Matthew A. Wills, "Phylogenetic position of Diania challenged," *Nature* 476 (August 11, 2011): E1; David A. Legg, Xiaoya Ma, Joanna M. Wolfe, Javier Ortega-Hernández, Gregory D. Edgecombe, and Mark D. Sutton, "Lobopodian Phylogeny Reanalyzed," *Nature* 476 (2011): E2; Jianni Liu, Michael Steiner, Jason A. Dunlop, Helmut Keupp, Degan Shu, Qiang Ou, JianHan, Zhifei Zhang, and Xingliang Zhang, "Liu et al. Reply," *Nature* 476 (2011): e3-e4.

34. Matzke, "Meyer's Hopeless Monster, Part II."

35. For example, Matzke obscures the many doubts about whether the legs of lobopods, or the eyes and head of anomalocaridids, are in fact homologous to (and therefore help explain the evolution of) the legs and eyes of arthropods. See Legg et al., "Cambrian bivalved arthropod reveals origin of arthrodization"; Edgecombe and Legg, "The Arthropod Fossil Record"; Liu et al., "Liu et al. reply," e3-e-4; Legg et al., "Lobopodian phylogeny reanalyzed," E2; Douglas Erwin and James Valentine, *The Cambrian Explosion: The Construction of Animal Biodiversity* (Greenwood Village, CO: Roberts and Company, 2013), 202; Edgecombe, "Arthropod phylogeny"; Stefan Richter, Martin Stein, Thomas Frace, and Nikolaus U. Szucsich, "The Arthropod Head," in *Arthropod Biology and Evolution: Molecules, Development, Morphology*, edited by A. Minelli et al. (Berlin: Springer-Verlag, 2013), 223–40. An organism cannot be transitional or intermediate to true arthropods if its features aren't homologous to arthropods. But if the legless anomalocaridids are more closely related to arthropods, and the arthropod head and eyes evolved first, then the legs of lobopods cannot be homologous to arthropod legs. Conversely, if the eyeless and headless lobopods are more closely related to arthropods, and arthropod legs evolved first, then the eyes and head of anomalocaridids cannot be homologous to arthropod eyes or heads. Pick either option, and you're faced with a situation that requires evolutionary loss of key arthropod traits, and at least one of Matzke's so-called transitional forms can't be transitional. As one recent *Nature* paper puts it, one of the "puzzles of stem-group arthropod evolution" is "the absence of [arthropod-like] trunk limbs in dinocaridids (*Anomalocaris*, etc.)," even though *Anomalocaris* has "a more arthropod-like head region" than lobopods, which lack the eyes or head of arthropods, but have arthropod-like "jointed trunk appendages." Liu et al., "Liu et al. reply," e3–e-4. In short, Matzke's clean evolutionary grade of intermediates leading to arthropods does not exist.

36. Erwin and Valentine, *The Cambrian Explosion*, 195, 202.

37. Edgecombe, "Arthropod phylogeny."

38. Legg et al., "Cambrian bivalved arthropod reveals origin of arthrodization."

39. Brysse, "From weird wonders to stem lineages," 306.

40. Ibid. (emphasis added).

41. Ibid., 311.

42. Ibid., 312 .

43. Henry Gee, *In Search of Deep Time: Beyond the Fossil Record to a New History of Life* (New York: Free Press, 2000), 151.

5.

A Graduate Student Writes

David Berlinski

Nick Matzke has written a critique of Stephen Meyer's *Darwin's Doubt*. Having for years defended Darwin's theory as an employee of the National Center for Science Education, he has determined to learn something about the theory as a graduate student at the University of California, an undertaking in the right spirit but the wrong order. Would that he had done things the other way around. His animadversions are written with all of the ebullience of a man sure enough of his conclusions not to worry overmuch about his arguments. They are wrong in the small, wrong in the large, and wrong all around. A pity. The Darwinian establishment is hardly without resources of its own, and had Matzke devoted more thought to his critique, he might have spared us the embarrassment of improving his arguments before rejecting his conclusions.

Darwin's Doubt advances three theses: First, that the Cambrian explosion is a real event; second, that the Cambrian explosion has not been explained by neo-Darwinian or other evolutionary mechanisms; and, third, that the Cambrian explosion, is best explained by an inference to intelligent design. Matzke in his critique of Meyer, concentrates on the first of these theses, barely mentions the second, and fails to engage Meyer's arguments for the third.

Darwin's Doubt makes its case for the reality of the Cambrian explosion chiefly, but not entirely, on the basis of the fossil record. Paleontology has pride of place. It is where the bodies are. Representatives of twenty-three of the roughly twenty-seven fossilized animal phyla, and the roughly thirty-six animal phyla overall, are present in the Cambrian fossil record. Twenty of these twenty-three major groups make their appearance with no discernible ancestral forms in either earlier Cambrian or Precambrian strata. Representatives of the remaining three or so animal phyla originate in the late Precambrian, but they do so as abruptly as the animals that appeared first in the Cambrian. Moreover, these late Precambrian animals lack clear affinities with the representatives of the twenty or so phyla that first appear in the Cambrian.

An account of their appearance must logically be focused on either the earliest part of the Cambrian or the Precambrian. Where else to look? But in looking there, Meyer argues, there is nothing much to see. Not *nothing*, of course. The well-known sequence that begins with the acritarchs and gutters into the small shelly fauna is an example, one in which Matzke invests his hopes without sufficiently hedging his bets. "The earliest identifiable representatives of Cambrian 'phyla,'" Matzke writes, with the twitch of misplaced quotation marks around a word that does not need them, "don't occur until millions of years after the small shelly fauna have been diversifying."[1]

But while the small shelly fauna offer something to see, they reveal nothing of interest. No paleontologist believes that some small shelly fauna are ancestral to *all* the Cambrian phyla. Trilobites are an example. These strange and complex creatures, their eyes staring hypnotically, appear during the early Cambrian (the Atdabanian stage). Having quite obviously gotten to where they appear in the fossil record, how did they get there? One speculative scenario runs from Precambrian bilaterians or arthropods to the Cambrian arachnomorphs, and then to the trilobites, the arrow of affirmative action in Lin et al. 2006 going from *Parvancorina* to *Skania sundbergi* and then wandering to *Primicaris larvaformis*.[2]

"What is often missed," Matzke argues, "is that deposits like the Chenjiang [sic] have dozens and dozens of trilobite-like and arthropod-like organisms." There follows a burst of exuberant thunder: "These are transitional forms!" Matzke is persuaded that whatever is trilobite-like must be trilobite-lite, and so ancestral to the trilobites themselves. The party line is otherwise:

> Early trilobites show all the features of the trilobite group as a whole; there do not seem to be any transitional or ancestral forms showing or combining the features of trilobites with other groups (e.g. early arthropods). Morphological similarities between trilobites and early arthropod-like creatures such as *Spriggina, Parvancorina*, and other "trilobitomorphs" of the Ediacaran period of the Precambrian are ambiguous enough to make detailed analysis of their ancestry far from compelling.[3]

If his natural allies in the great cause have refrained from supporting his conclusions, Matzke is prepared to advance them anyway, a policy commanding our admiration if only for its foolhardiness: "All of this is pretty good evidence," Matzke writes, "for the basic idea that the Cambrian 'Explosion' is really the radiation of simple bilaterian worms into more complex worms, and that this took something like 30 million years just to get to the most primitive forms that are clearly related to one or another living crown 'phyla,' and occurred in many stages, instead of all at once."[4]

This is a view championed by Matzke in defiant isolation. The University of California's Museum of Paleontology makes the obvious case to the contrary:

> When the fossil record is scrutinized closely, it turns out that the fastest growth in the number of major new animal groups took place during the as-yet-unnamed second and third stages (generally known as the Tommotian and Atdabanian stages) of the early Cambrian, a period of about 13 million years. In that time, the first undoubted fossil annelids, arthropods, brachiopods, echinoderms, molluscs, onychophorans, poriferans, and priapulids show up in rocks all over the world.[5]

Matzke is pursuing his PhD at the University of California. He is apparently indisposed to visiting museums.

Notes

1. Nick Matzke, "Meyer's Hopeless Monster Part II," *Panda's Thumb,* June 19, 2013, http://pandasthumb.org/archives/2013/06/meyers-hopeless-2.html.

2. J. P. Lin, et al., "A *Parvancorina*-like arthropod from the Cambrian of South China." *Historical Biology* 18, no. 1 (2006): 33–45.

3. From *Wikipedia*, a Party Organ, and hardly a source our side is disposed to champion. Why stop there? See P. Jell, "Phylogeny of Early Cambrian trilobites," in Philip D. Lane et al., *Trilobites and Their Relatives: Contributions from the Third International Conference, Oxford 2001*, vol. 70 (London: Palaeontological Association, 2003), 45–57.

4. Matzke, "Meyer's Hopeless Monster Part II."

5. "The Cambrian Period," *University of California Museum of Paleontology,* July 6, 2011, http://www.ucmp.berkeley.edu/cambrian/cambrian.php.

6.

How "Sudden" Was the
Cambrian Explosion?

Casey Luskin

On June 19, the day after *Darwin's Doubt* was first available for purchase, Nick Matzke published a 9400-word "review" of the book in which it appears that he tried to anticipate many of Stephen Meyer's arguments. Unfortunately, he often either guessed wrong as to what Meyer would say or—assuming he actually read the book as he claims—misread many of Meyer's specific claims. Matzke repeatedly misquoted Meyer, at one point claiming he referred to the Cambrian explosion as "instantaneous," when Meyer nowhere makes that claim. Indeed, Matzke faulted Meyer for not recognizing that the Cambrian explosion "was not really 'instantaneous' nor particularly 'sudden.'" Oddly, he also criticized Meyer for not recognizing that the Cambrian explosion "took at least 30 million years"—despite expert opinion showing it was far shorter.[1]

After Matzke published his review, *The New Yorker* reviewed Meyer's book. Gareth Cook, the science writer who wrote the piece, relied heavily on Matzke's critical evaluation, even though Matzke was at the time a graduate student and not an established Cambrian expert. Cook uncritically recycled Matzke's claim that the Cambrian explosion took "many tens of millions of years," even saying that the main problem with *Darwin's Doubt* is that Meyer failed to recognize this alleged fact.[2]

So, was Matzke right about the length of the Cambrian explosion? In fact, Matzke's preemptive—or hastily written—review not only misrepresented Meyer's view; it also misrepresented the length and character of the Cambrian explosion, as numerous authoritative peer-reviewed scientific sources on the subject clearly show.

Before going on, let's briefly look first at what Meyer says. First, Meyer does not equate the Cambrian explosion with the *entire* radiation—as most Cambrian experts also do not. By "radiation" here I mean the period of time in which *all* the new phyla, classes, orders that first arose during the Cambrian apparently did so. Instead, he equates the Cambrian explosion with the most explosive period of the Cambrian radiation (as most Cambrian experts do) in which the vast majority of the higher taxa arose. He asserts specifically that the re-dating of critical Cambrian strata in 1993 established that the first appearance of the majority of the Cambrian phyla and classes took place within a 10 million year period—a period Meyer does equate with "the explosion of novel Cambrian animal forms." (pp. 71–72) As he describes it, "these studies [i.e., radiometric analyses of zircon crystals in Siberian rocks] also suggested that the explosion of novel Cambrian animal forms" took about 10 million years. (p. 71)

In affirming this, however, Meyer offers a nice discussion of how different scientists may judge the duration of the Cambrian explosion differently, depending upon how they choose to define it and how many *separate* events they decide to include. (pp. 71–73) Thus, Meyer notes that if paleontologists decide to include as part of the Cambrian explosion (a) the origin of the Ediacaran organisms in the late Precambrian, *and* (b) the small shelly fossils at the base of the Cambrian, *and* (c) the main pulse of morphological innovation in the early Cambrian, *and* (d) subsequent diversification events right up until the end of the Cambrian period, they might claim that the Cambrian explosion lasted nearly 80 million years, as, for example, geologist Donald Prothero does (a point

Meyer also notes in his book). Nick Matzke appears to include in the Cambrian explosion everything from the appearance of the small shelly fossils at the base of the Cambrian (541 million years ago) to the main pulse of morphological innovation (530–520 million years ago) to events in the late Cambrian (about 512–505 million years ago).

In any case, Meyer recognizes the conventional and somewhat subjective nature of attempts to define and delimit "the Cambrian explosion." He nevertheless accepts a 10-million-year duration of the explosion itself, in keeping with the common judgment of numerous Cambrian experts about the length of time in which the vast majority of new phyla and classes arose—as I will document below. Yet, to circumvent issues of semantics and subjective definitions, Meyer focused his analysis on the problem of the origin of *novel* animal form, and, thus, the main or most explosive pulse of such "morphological innovation." This makes sense because the problem that Meyer ultimately addresses, and the problem that evolutionary biology must address, is that of building novel animal forms or body plans in the first place. Can the neo-Darwinian mechanism generate the amount of novel form and information that arises in the Cambrian period in the time allowed by the fossil record? By focusing his analysis on the main period of morphological innovation, Meyer defines clearly the most salient challenge posed to the adequacy of neo-Darwinian (and other evolutionary) mechanisms.

To establish the length of the most explosive period of innovation within the Cambrian explosion itself, Meyer cites the work of MIT geochronologist Samuel Bowring and his colleagues as well the work of another group led by Smithsonian paleontologist Douglas Erwin. The Bowring-led study showed that (in their words) the main "period of **exponential increase of diversification** lasted only 5 to 6 m.y." and is "unlikely to have exceeded 10 m.y."[3] Meyer explains:

> An analysis by MIT geochronologist Samuel Bowring has shown that the main pulse of Cambrian morphological innovation occurred in a sedimentary sequence spanning no more than 6 million years. Yet during this time representatives of at least sixteen completely novel phyla and about thirty classes first ap-

peared in the rock record. In a more recent paper using a slightly different dating scheme, Douglas Erwin and colleagues similarly show that thirteen new phyla appear in a roughly 6-million-year window.[4]

To see why Meyer made these claims, take a look first at Figure 6-1, which Bowring and his colleagues included in their definitive 1993 article, published in the journal *Science*.

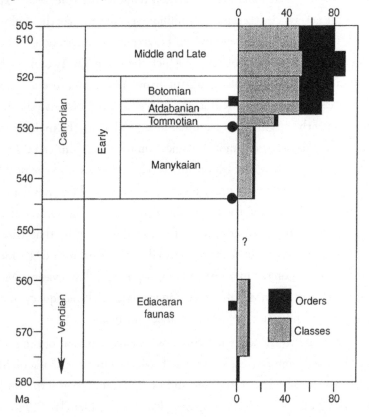

Figure 6-1. From Samuel A. Bowring, John P. Grotzinger, Clark E. Isachsen, Andrew H. Knoll, Shane M. Pelechaty, Peter Kolosov, "Calibrating Rates of Early Cambrian Evolution," *Science*, 261 (3 September 1993), 1293–1298. Reprinted with permission from AAAS.

In that article, they use radiometric methods to date the different stages of the Cambrian period, including the crucial Tommotian and Atdabanian stages in which the greatest number of new animal phyla and classes arise. Note that the so-called Manykaian stage of the Cambrian period lasts about 10–14 million years. Note also that the main pulse of morphological innovation didn't begin during this stage but rather during the Tommotian and Atdabanian—a period that they describe as taking between "5 to 10 million years," and in a more detailed passage as taking about 5–6 million years.

In Figure 6-1, the Tommotian and Atdabanian stages of the Cambrian period together span only about 5 million years, starting at about 530 and ending about 525 million years ago. Bowring's figure also depicts the total number of classes and orders present at any given time during the Cambrian period. The biggest increases in morphological innovation occur during the Tommotian and Atdabanian stages. Indeed, during this period the number of known *orders* nearly quadruples. Moreover, Bowring and his colleagues make clear that this period corresponds to the main pulse of Cambrian morphological innovation as measured by the number of new *phyla* and *classes* that first appear. They note that, while a few groups of animals do arise in the earliest Manykaian stage of the Cambrian, the most rapid period of "exponential increase of diversification," corresponding to the Tommotian and Atdabanian stages, "lasted only 5 to 6 m.y." They explain:

> [T]he initial (Manykaian) interval of slow diversification followed the ediacaran faunal epoch by no more than 20 million years (m.y.) and lasted approximately 14 m.y. **In contrast, if we accept the age of 525 Ma for the Atdabanian-Botomian boundary, then the Tommotian-Atdabanian period of exponential increase of diversification lasted only 5 to 6 m.y. In any event it is unlikely to have exceeded 10 m.y. Numbers of phyla, classes, orders, families, and genera all reached or approached their Cambrian peaks during the short Tommotian-Atdabanian interval.** For phyla and classes, most of the diversity known for the Phanerozoic [the eon of time since the Cambrian] as a whole differentiated by the end of the Atdabanian.[5]

In Chapter 3 (p. 73), Meyer also cites a 2011 paper by Douglas Erwin and several colleagues. Although Erwin et al. use slightly different starting and ending dates and different names for the stages of the Cambrian period, they too estimate that the most explosive stage took about

5–6 million years. Indeed, the supplemental documentation to their article shows 13 or 14 new phyla arising during "Stage 3" of the Cambrian period, a stage that corresponds to a narrow 5–6 million-year window (see Figure 3 in their article), just as Meyer wrote in *Darwin's Doubt*.

Erwin and his colleagues note that "most paleontologists favor a near literal reading of the fossil record, supporting a rapid (~25-million-year) evolutionary divergence of most animal clades near the base of the Cambrian"[6]—a duration a bit shorter than but close to the "at least 30 million years" given by Matzke. But here the authors are talking about not only the most explosive stage (Stage 3) or stages (Stages 2 and 3) of the Cambrian, but also Stage 1, which they and most experts usually exclude from "the Cambrian explosion."

Indeed, Erwin, writing more recently with James Valentine in their book *The Cambrian Explosion*, dates the Cambrian explosion to **"a geologically brief interval between about 530 to 520 Ma"**:

> [A] great variety and abundance of animal fossils appear in deposits dating from a **geologically brief interval between about 530 to 520 Ma,** early in the Cambrian period. During this time, nearly all the major living animal groups (phyla) that have skeletons first appeared as fossils (at least one appeared earlier). Surprisingly, a number of those localities have yielded fossils that preserve details of complex organs at the tissue level, such as eyes, guts, and appendages. In addition, several groups that were entirely soft-bodied and thus could be preserved only under unusual circumstances also first appear in those faunas. Because many of those fossils represent complex groups such as vertebrates (the subgroup of the phylum Chordata to which humans belong) and arthropods, it seems likely that all or nearly all the major phylum-level groups of living animals, including many small soft-bodied groups that we do not actually find as fossils, had appeared by the end of the early Cambrian. This **geologically abrupt** and spectacular record of early animal life is called the Cambrian explosion.[7]

Like many Cambrian experts, Erwin and Valentine focus their analysis on that part of the Cambrian radiation in which the greatest amount of morphological innovation arises—and define "the Cambrian explosion" accordingly. They believe that nearly the full breadth of Cambrian diversity arose in less than ten million years, writing: "the basic structure of Phanerozoic ecosystems had been achieved **within at most 10 million years** after the onset of bilaterian diversification."[8]

Moreover, many other Cambrian experts focus on precisely this period of the origin of maximum morphological novelty in their discussion (and definition) of the Cambrian explosion. They define the Cambrian explosion as an event that encompassed about (or even less than) 10 million years just as Meyer does, not one that took "at least 30 million years" as Matzke claims. For example:

Prominent paleontologist Robert Carroll stated in *Trends in Ecology and Evolution* that the Cambrian explosion took less than ten million years:

> The most conspicuous event in metazoan evolution was the dramatic origin of major new structures and body plans documented by the Cambrian explosion. Until 530 million years ago, multicellular animals consisted primarily of simple, soft-bodied forms, most of which have been identified from the fossil record as cnidarians and sponges. Then, **within less than 10 million years**, almost all of the advanced phyla appeared, including echinoderms, chordates, annelids, brachiopods, molluscs and a host of arthropods. The extreme speed of anatomical change and adaptive radiation during this brief time period requires explanations that go beyond those proposed for the evolution of species within the modern biota.[9]

An article in the journal *Development* by Erwin, Valentine and David Jablonski explains that:

> The Cambrian explosion is named for the **geologically sudden** appearance of numerous metazoan body plans (many of living phyla) **between about 530 and 520 million years ago**, only 1.7% of the duration of the fossil record of animals.[10]

Another article in a major evolution journal states that:

> … recent geological investigations suggest that the Cambrian explosion may have occurred **within a period of only 5–10 million years**.[11]

A paper in *BioEssays* states:

> Because of the sudden appearance of a near complete diversity of animal body plans in the fossil record around **530–520 million years ago,** this diversification is commonly referred to as the "Cambrian explosion."[12]

Another paper by the eminent biologist Susumu Ohno states:

> ... this Cambrian explosion, during which nearly all the extant animal phyla have emerged, was of an **astonishingly short duration, lasting only 6–10 million years.**[13]

A paper by Andrew R. Parker of the Department of Zoology at the Natural History Museum in London states:

> The Cambrian explosion, or Big Bang in animal evolution, was the most dramatic event in the history of life on Earth. During this blink of an eye in such history, most phyla found today evolved their first hard parts and distinct shapes at the same time. In other words, it is the event where animals suddenly took on very different appearances, in the form they exist today. The event itself, however, occupied only a small part of the Cambrian period, somewhere between **520 and 515 Ma.** Prior to this, there were only three animal phyla with the type of external shapes they still possess today. Yet in a geological instant later there were at least several more—and perhaps most—of the phyla known today.[14]

Even a 2007 paper in *Journal of College Science Teaching,* authored by zoologist Thomas Gregg (who criticizes intelligent design in the article), states:

> The Cambrian explosion is the appearance of several dozen fossilized species with different body plans over a period of **5–15 million years** during the Cambrian period.[15]

In case you didn't notice, none of these authorities are saying the Cambrian explosion "took at least 30 million years."

Matzke does cite one paper when attempting to justify his claim that the Cambrian explosion "took at least 30 million years, and was not really 'instantaneous' nor particularly 'sudden.'" But that source—a 2005 paper in *Paleobiology* by Kevin J. Peterson, Mark A. McPeek, and David A. D. Evans—does not place exact numbers on the time scale of the Cambrian explosion, so it doesn't help Matzke's case much. Indeed, a close analysis of the figure[16] Matzke posts from that paper shows that it too reveals a rapid pulse of diversification in the mid-early Cambrian. Moreover, two of those three authors directly contradicted Matzke's thesis about the length of the Cambrian explosion in a paper in *BioEssays*, published four years later:

> Part of the intrigue with the Cambrian explosion is that numerous animal phyla with very distinct body plans arrive on the scene in a **geological blink of the eye**, with little or no warning of what is to come in rocks that predate this interval of time. The **abruptness** of the transition between the "Precambrian" and the Cambrian was apparent right at the outset of our science with the publication of Murchison's The Silurian System, a treatise that paradoxically set forth the research agenda for numerous paleontologists—in addition to serving as perennial fodder for creationists. The reasoning is simple—as explained on an intelligent-design t-shirt.
>
> > Fact: Forty phyla of complex animals **suddenly appear** in the fossil record, **no forerunners, no transitional forms** leading to them; "a major mystery," a "challenge." The Theory of Evolution—exploded again (idofcourse. com).
>
> Although we would dispute the numbers, and aside from the last line, **there is not much here that we would disagree with**. Indeed, many of Darwin's contemporaries shared these sentiments, and we assume—if Victorian fashion dictated—that they would have worn this same t-shirt with pride.[17]

Matzke appears unaware of what the very authorities he cites have said about the length of the Cambrian explosion.

Indeed, unquestionably, many senior Cambrian paleontologists and other established Cambrian experts contradict Matzke's claim about the length of the Cambrian explosion. Of course, Matzke is free to define the Cambrian explosion in whatever idiosyncratic way he chooses. However, by defining it as a series of separate events in the fossil record spanning "at least 30 million years" he not only introduces confusion about a term with a relatively stable meaning in paleontology, but he diverts attention from the crucial problem of explaining the most explosive appearance of evolutionary and morphological novelty that the phrase "Cambrian explosion" has commonly been used to describe.

What about Matzke's claim that Meyer should not have referred to the event as geologically "sudden"? We have already seen that Valentine, Jablonski, and Erwin called the Cambrian explosion "geologically sudden." As it turns out, many other authors in the technical literature have used that exact terminology to describe the Cambrian explosion:

+ "Nobody seriously doubts that the **sudden appearance** in the fossil record of numerous marine animal groups of both familiar and enigmatic type close to the base of the Cambrian reflects one of the important events in the history of the biosphere."[18]

+ "Beautifully preserved organisms from the Lower Cambrian Maotianshan Shale in central Yunnan, southern China, document the **sudden appearance** of diverse metazoan body plans at phylum or subphylum levels, which were either short-lived or have continued to the present day."[19]

+ "the **sudden expansion** in phyla of the Cambrian explosion"[20]

+ "Most of the animal phyla that are represented in the fossil record first appear, **fully formed** and identifiable as to their phylum, in the Cambrian.... The fossil record is therefore of no help with respect to understanding the origin and early diversification of the various animal phyla..."[21]

+ "the **sudden appearance** of a near complete diversity of animal body plans in the fossil record around 530–520 million years ago."[22]

+ "the profound morphological gaps among the major groups, set against the background of **sudden appearances** in the fossil record of many novel taxa and the absence of easily recognizable transitional forms."[23]

+ "Darwin recognized that the **sudden appearance** of animal fossils in the Cambrian posed a problem for his theory of natural selection. ... Recent geochronological studies have reinforced the impression of a 'big bang of animal evolution' by narrowing the temporal window of apparent divergences to just a few million years."[24]

+ "The apparently **sudden origin** of animal phyla has contributed to the view that phyla represent a fundamental level of organization."[25]

+ "The fossil record of metazoa shows a **sudden expansion** at around 550–530 million years ago."[26]

+ "This paucity of metazoan fossils in the strata of Earth is broken by the **sudden appearance** of highly developed metazoan fossils in the Cambrian, a pattern colloquially referred to as the Cambrian evolutionary 'explosion.'"[27]

+ "[T]he fossil record displays the **sudden appearance** of intracellular detail and the 32 phyla."[28]

+ "The Cambrian explosion in animal evolution during which all the diverse body plans appear to have emerged almost in a **geological instant** is a highly publicized enigma."[29]

+ "At the beginning of the Cambrian, however, life took a **sudden** turn toward the complex. In a few million years—the equivalent of a **geological instant**—an ark's worth of sophisticated body types filled the seas. This biological **burst**, dubbed the Cambrian explosion, produced the first skeletons and hard shells, antennae and legs, joints and jaws. It set the evolutionary stage for all that followed by giving rise to most of the major phyla known on Earth today. Even our own chordate ancestors got their start during this long-past era."[30]

So, again, Matzke's claims stand at odds with the technical literature in a field he purports to represent. At the very least, Meyer seems fully justified in calling the Cambrian explosion "sudden" because so many other authorities use that same term. It's too bad *The New Yorker*, once legendary for meticulous fact-checking, didn't dig a little deeper but instead relied on Matzke's claims, which have turned out to be incorrect.

Finally, note that even if Matzke's 30-million-year figure were correct, it would not help his case. As Meyer shows in Chapter 10 and Chapter 12 of *Darwin's Doubt*, the extreme rarity of genes and proteins in sequence space means that even thirty million years is not nearly enough time to give the neo-Darwinian mechanism a realistic opportunity to generate a new gene or protein—let alone a new form of animal life. Further, as he shows in Chapter 12, the calculated waiting times using the standard principles of population genetics for the occurrence of just a few (three or more) coordinated mutations vastly exceed 30 million years. In his review, Matzke summarily dismissed these arguments, neither engaging nor rebutting them.

Notes

1. Nick Matzke, "Meyer's Hopeless Monster Part II," *Panda's Thumb*, June 19, 2013, http://pandasthumb.org/archives/2013/06/meyers-hopeless-2.html.

2. Gareth Cook, "Doubting 'Darwin's Doubt,'" *The New Yorker*, (July 2, 2013), http://www.newyorker.com/tech/elements/doubting-darwins-doubt. For more on Cook's article see: David Klinghoffer, "From *The New Yorker*, Backhanded Compliments for *Darwin's Doubt*," *Evolution News & Views*, July 2, 2013, http://www.evolutionnews.org/2013/07/from_the_new_yo074041.html.

3. S. A. Bowring et al., "Calibrating rates of early Cambrian evolution," *Science* 261, no. 5126 (3 September 1993): 1293–1298, 1297, http://www.sciencemag.org/content/261/5126/1293.abstract. Emphasis added.

4. Stephen C. Meyer, *Darwin's Doubt: The Explosive Origin of Animal Life and the Case for Intelligent Design* (New York: HarperOne, 2013), 73.

5. Bowring et al., "Calibrating dates," 1297. Emphasis added.

6. Douglas H. Erwin et al., "The Cambrian Conundrum: Early Divergence and Later Ecological Success in the Early History of Animals," *Science* 334 (2011): 1991–97, http://www.sciencemag.org/content/334/6059/1091.full.

7. Douglas Erwin and James Valentine, *The Cambrian Explosion: The Construction of Animal Biodiversity* (Greenwood Village, CO: Roberts and Company, 2013), 5. Emphasis added.

8. Ibid., 266. Emphasis added.

9. Robert L. Carroll, "Towards a new evolutionary synthesis," *Trends in Ecology and Evolution* 15 (January 2000): 27–32, http://www.cell.com/trends/ecology-evolution/abstract/S0169-5347(99)01743-7. Emphasis added.

10. James W. Valentine, David Jablonski, and Douglas H. Erwin, "Fossils, molecules and embryos: new perspectives on the Cambrian explosion," *Development* 126 (1999): 851–859, http://dev.biologists.org/content/126/5/851.full.pdf. Emphases added.

11. Michael A. Bell, "Origin of the metazoan phyla: Cambrian explosion or proterozoic slow burn," *Trends in Ecology and Evolution* 12 (January 1, 1997): 1–2, http://www.ncbi.nlm.nih.gov/pubmed/21237949. Emphasis added.

12. Tanya Vavouri and Ben Lehner, "Conserved noncoding elements and the evolution of animal body plans," *BioEssays* 31 (2009): 727–735, http://www.ncbi.nlm.nih.gov/pubmed/19492354. Emphasis added.

13. Susumu Ohno, "The notion of the Cambrian pananimalia genome," *Proceedings of the National Academy of Sciences, USA* 93 (August, 1996): 8475–8478, http://www.pnas.org/content/93/16/8475.full.pdf. Emphasis added.

14. Andrew R. Parker, "On the origin of optics," *Optics & Laser Technology*, 43 (2011): 323–329. Emphasis added.

15. Thomas Gregg, "Intelligent Design: Jonathan Wells and the Tree of Life," *Journal of College Science Teaching* 36, no. 7 (July/August, 2007): 10. Emphasis added.

16. "Figure 2. Tempo of early animal evolution placed into the geologic context of the Neoproterozoid/Cambrian transaction," *Panda's Thumb*, http://www.pandasthumb.org/archives/cambrian.JPG.

17. Kevin J. Peterson, Michael R. Dietrich, and Mark A. McPeek, "MicroRNAs and metazoan macroevolution: insights into canalization, complexity, and the Cambrian explosion," *BioEssays*, 31 no. 7 (July 2009): 736–747, http://onlinelibrary.wiley.com/doi/10.1002/bies.200900033/abstract. Emphases added.

18. R. A. Fortey, D. E. G. Briggs, and M. A. Wills, "The Cambrian evolutionary 'explosion': decoupling cladogenesis from morphological disparity," *Biological Journal of the Linnean Society* 57, no. 1 (January 1996): 13–33, http://onlinelibrary.wiley.com/doi/10.1111/j.1095-8312.1996.tb01693.x/abstract. Emphasis added.

19. Jun-Yuan Chen, "The sudden appearance of diverse animal body plans during the Cambrian explosion," *International Journal of Developmental Biology* 53 (2009): 733–51, http://www.ijdb.ehu.es/web/paper/072513cj/the-sudden-appearance-of-diverse-animal-body-plans-during-the-cambrian-explosion. Emphases added.

20. Lynn Helena Caporale, "Putting together the pieces: evolutionary mechanisms at work within genomes," *BioEssays* 31, no. 7 (July 2009): 700–702, http://onlinelibrary.wiley.com/doi/10.1002/bies.200900067/abstract. Emphasis added.

21. R. S. K. Barnes et al., *The Invertebrates: A New Synthesis*, 3rd ed. (Malden, MA: Blackwell Scientific Publications, 2001): 9–10. Emphasis added.

22. T. Vavouri and B. Lehner, "Conserved noncoding elements and the evolution of animal body plans," *BioEssays* 31, no 7 (July 31, 2009): 727–35, http://onlinelibrary.wiley.com/doi/10.1002/bies.200900014/abstract. Emphasis added.

23. Richard K. Grosberg, "Out on a Limb: Arthropod Origins," *Science* 250, no. 4981 (2 November 1990): 632–633, http://www.sciencemag.org/content/250/4981/632. Emphasis added.

24. Gregory A. Wray, Jeffrey S. Levinton, and Leo H. Shapiro, "Molecular Evidence for Deep Precambrian Divergences," *Science* 274, no. 5287 (October 25, 1996): 568–573, http://www.sciencemag.org/content/274/5287/568. Emphasis added.

25. Lindell Bromham, "What can DNA Tell us About the Cambrian Explosion?," *Integrative and Comparative Biology* 43, no. 1 (February 2003): 148–156, http://icb.oxfordjournals.org/content/43/1/148. Emphasis added.

26. Andrew M. Sugden, "Tracing the Explosion to its Roots," *Science* 288, no. 5468 (May 12, 2000): 929. Emphasis added.

27. Christopher W. Wheat and Niklas Wahlberg, "Phylogenomic Insights into the Cambrian Explosion, the Colonization of Land and the Evolution of Flight in Arthropoda," *Systematic Biology* 62, no. 1 (January 1, 2013), 93–109, http://sysbio. oxfordjournals.org/content/62/1/93. Emphasis added.

28. Michael A. Crawford et al., "A quantum theory for the irreplaceable role of docosahexaenoic acid in neural cell signalling throughout evolution," *Prostaglandins, Leukotrienes and Essential Fatty Acids* 88, no. 1 (January 2013): 5–13, http://www.plefa. com/article/S0952-3278(12)00147-0/abstract. Emphasis added.

29. Eugene V. Koonin, "The Biological Big Bang model for the major transitions in evolution," *Biology Direct* 2, no. 21 (August 20, 2007), http://www.biologydirect.com/ content/2/1/21. Emphasis added.

30. Richard Monastersky, "The First Monsters: Long before Sharks, Anomalocaris Ruled the Seas," *Science News* 146, no. 9 (August 27 1994): 138–39. Emphasis added.

7.

A One-Man Clade

David Berlinski

The problems associated with the biological character problem [cladistics] are so complex and multifaceted and this issue is so conceptually immature that any single author's account is doomed to be too narrow and lopsided to be of much use.[1]

— Günter Wagner

H AD STEPHEN MEYER BETTER APPRECIATED THE TOOLS OF MOD-ern cladistics, Nick Matzke believes, he would not have drawn the conclusions that he did in his book *Darwin's Doubt*, or argued as he had. Meyer is in this regard hardly alone. It would seem that Stephen Jay Gould was just slightly too thick to have appreciated, and the eminent paleontologist James Valentine just slightly too old to have acquired, the methods that Matzke, writing at *Panda's Thumb*, is disposed to champion.[2] Should Valentine be appointed to Matzke's dissertation committee at UC Berkeley, we at Discovery Institute will be pleased to offer uninterrupted prayers on his behalf. We can offer no assurance of success, of course, but then again, when it comes to cladistic methods, neither can Matzke.

Why, Matzke wonders, did Stephen Meyer not include within his book cladograms such as those he himself displays in his critique, one due to Brysse, the other to Legg? He is in asking this question in full Matzke mode: sleek with satisfaction. Meyer may well have refrained from including *these* cladograms because they are topologically in conflict, and display virtually no agreement with one another. Matzke's inability to discern what is directly beneath his nose is hardly evidence of his own competence in cladistic analysis.[3]

It was the German entomologist Willi Hennig who with the publication of *Grundzüge einer Theorie der phylogenetischen Systematik* introduced biologists to his scheme of classification.[4] Matzke is surely right to remark that drawing up character sets is a detailed and tedious business. But so is the work involved in alphabetizing the names of all the residents in Moscow in 1937. It is in either case no great recommendation. The great merit of cladistic analysis is just the work that it makes for cladistic analysts.[5] Like so much in Darwinian biology, it is a gift that keeps on giving.

A cladistic system expresses a complicated jumble of assumptions and definitions, these expressed most often in the baroque and oddly beautiful vocabulary of Greek and Latin technical terms. No taxonomist with access to words such as *paraphyletic, plesiomorphy,* or *synapomorphy* would ever be satisfied by a description of *Anomalocaris* as some bug-eyed monster shrimp. Not me, for sure. Assumptions and definitions in cladistics sheathe a sturdy but simple skeleton, nothing more than a graph, lines connected to points in the plane. The blunt, no-nonsense language of graph theory is quite sufficient. A graph $G = <V, E>$ is a collection of vertices and edges. A given vertex may be either an intersection or a terminal point of a graph. A Steiner tree is a graph spanning its terminal points. Although Steiner trees are designed to be unobtrusive, like any skeleton, they make demands all their own, most obviously because they are finite and discrete.

The cladistic classification of the living and the dead proceeds by means of a character matrix, one whose elements are bright but isolated morphological or genetic bits. Fingers are ideal. They can be counted. No person (typically) has more than five of them. And fingers are discrete. No one need wonder where one begins and the other ends, a point not lost on short-tempered motorists. When character sets are expressed as graphs, the result is a cladogram.

Five taxa in all: **A, B, C, D,** and **E,** individuals, species, or whole slobbering groups of them. Assume that this is so. And four characters: **1, 2, 3,** and **4.** A four by five matrix suffices to display the distribution of characters, with 1 indicating that a character is present in a taxon, and 0 that it is not:

	1	2	3	4
A	0	0	0	1
B	0	0	1	1
C	1	0	1	1
D	1	1	1	1
E	1	1	1	1

The translation of a character matrix into a cladogram can in this simple case be done by hand. Anything more complicated requires a computer program. The clade of cladists and the clade of computer programmers are on the best of terms. One clade washes the other, as professionals so often observe. Terminal points in a cladogram are occupied by the names of taxa, and vertices by their characters, as in Figure 7-1.

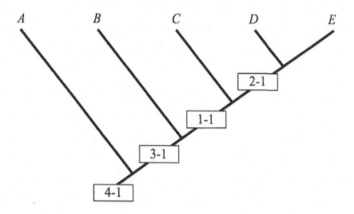

Figure 7-1. Illustration: R. Sternberg.

In this cladogram, the boxed labels (4-1, 3-1, 1-1, 2-1) indicate when the different characters are supposed to have arisen in the phylogenetic history of this group of organisms (A, B, C, D, and E). The first number of each label represents a specific character, while the second number ("1") indicates that the character was gained at that point in phylogenetic history. The placement of a label before one or more taxa implies that the identified character is present in all following taxa unless it is subsequently lost (however, there are no examples of character loss in this diagram). For example, in this diagram, the placement of label 4-1 shows that character 4 is present in taxa A, B, C, D, and E. Similarly, the placement of label 2-1 shows that character trait 2 is present only in taxa D and E.

Aber sei vorsichtig, as Willi Hennig himself might have said, and, no doubt, would have said had he read Matzke's critique. It is remarkably difficult to read a cladogram without reading something into it, more than the graph conveys, a bit of Darwinian doggerel most often. **A, B, C, D** and **E** are labels marking points in the plane; the taxa that they designate are found in nature. There is a difference. That **A** is to the left of **B** is a fact about graphs and labels. It makes no sense to say of two taxa that one is to the left of the other. Very few taxonomists are known widely to confuse their left and their right hands—no more than one or two. This is reassuring. That **B** is *between* **A** and **C** is otherwise. It is tempting. It is tempting precisely because it invites the taxonomist to undertake an inference from the premise that **B** is between **A** and **C** to the conclusion that **B** is somehow a descendant of **A,** an ancestor of **C**.

A cladogram does not by itself justify anything of the sort: The inference remains a non-starter because it exhibits a non-sequitur. "Evolution[ary theory] is not a necessary assumption of cladistics," the biologist A. V. Z. Brower once remarked.[6] Neither is it sufficient, I would add. A cladogram is one way of depicting the information resident in a character matrix, and given the open-ended relationship between a matrix and its depiction in a cladogram, it is by no means unique. Intermediates that are clear as sunshine in a given cladogram disappear into darkness when part of the cladogram is rotated.

There is thus Figure 7-2.

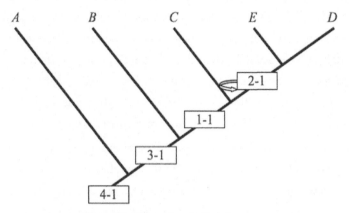

Figure 7-2. Illustration: R. Sternberg.

Or even Figure 7-3.

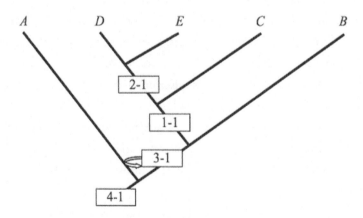

Figure 7-3. Illustration: R. Sternberg.

The cladograms in Figures 7-2 and 7-3 preserve the hierarchical structure of Figure 7-1, but they fail notably to keep intermediate taxa where they once belonged. Rotations preserve some of the structure of the original, but not all of it. Similarity in structure may well be an assumption governing the identity of various cladograms; but as these examples show, structural similarity along one dimension does *not* preserve structural similarity along another. A cladist championing Figure 7-1 in his PhD dissertation is apt to see intermediates in the fossil record that his colleagues, and so his competitors, regard as nothing more than so many graph-theoretic artifacts. Never mind. Championing Figures 7-1 and 7-2 in *their* dissertations, they have artifacts all their own to exhibit or to hide. It is a very good thing that these people are seldom armed and rarely dangerous. If this is so, how, then, to define *transitional* forms? If no definition is possible, then the relevance of cladistic analysis to Darwinian biology might be more limited than often thought. It is in any case an obvious question to ask, the more so when assessing a book calling attention to the *absence* of transitional forms in the Cambrian era.[7]

Matzke acknowledges the point without grasping its meaning: "[P]hylogenetic methods as they exist now," he writes, "can only rigorously detect sister-group relationships, not direct ancestry, and, crucially, … this is neither a significant flaw, nor any sort of challenge to common ancestry, nor any sort of evidence against evolution." But there can be no sisters without parents, and if cladistic analysis cannot detect their now mythical ancestors, it is hard to see what is obtained by calling them sisters. No challenge to common ancestry? Fine. *But no support for common ancestry either.* Questions of ancestry go beyond every cladistic system of classification. It follows that questions with respect to the ancestry of various Cambrian phyla cannot be resolved by *any* cladistic system of classification, however its characters are defined. We are now traveling in all the old familiar circles. The claim made by *Darwin's Doubt* is that with respect to the ancestors of those Cambrian phyla, there is nothing there.

The relationship between cladistics and Darwin's theory of evolution is thus one of independent origin but convergent confusion. "Phylogenetic systematics," the entomologist Michael Schmitt remarks, "relies on the theory of evolution." To the extent that the theory of evolution relies on phylogenetic systematics, the disciplines resemble two biologists dropped from a great height and clutching at one another in mid-air.

Tight fit, major fail.[8]

No wonder that Schmidt is eager to affirm that "phylogenetics does not claim to prove or explain evolution whatsoever."[9] If this is so, a skeptic might be excused for asking what it does prove or explain?

The graphs and trees of cladistic analysis are when examined statistically capable of recording strong signals. Matzke is right to say this.[10] He is himself so alert to them that he resembles an old-fashioned commodore peering beyond mist and mizzen and hoping to see flags. While those signals may be strong, it is often unclear what they are signaling. We may imagine the world's living cladists sorted by a complex character matrix, one involving graduate degrees, publications, tenure, citations, beard length, night sweats, beady eyes, prison records, and trans-gendered identity. Their crown group comprises the living cladists together with their last common ancestor and all of his descendants. Cladistic analysis indicates, I am at liberty to disclose, that Willi Hennig is the last common ancestor of all living cladists, and so an *Ur-Mensch*, another reason for the respect in which he is held. Having branched off early, Sokal and Sneath make for a sister group. Extinct stem groups may be seen as well, tracing *their* ancestry to an *unter-Mensch*, as German cladists say, one lost in the fog of time and more basal, if not more base, than Willi Hennig himself.

A cladogram *of* cladists is no different from the real thing, a cladogram *by* cladists. But Willi Hennig is *not* the last common ancestor of contemporary cladists, the gorgeous apparatus of cladistics notwithstanding, and no cladistic *unter-Mensch* ever existed. Common characters and common descent are not the same thing.

Cladistic methods thus suggest a number of reservations. Character matrices are the method's heart and soul, ineliminable in practice and theory. It is precisely because character matrices are finite and discrete that cladists believe that they have on hand a body of data that they can master. "Cladistics breaks up the bodyplan characters," as Matzke observes, "and shows the basic steps they evolved in, and also which parts of the 'bodyplan' are actually shared with other phyla."[11] This is *precisely* what cladistics does, but what it does is at least open to the suspicion that when it comes to these issues, cladistic analysis is driven more by what cladists can do than what they should do. "No experienced naturalist," Stephen Jay Gould remarked, "could ever fully espouse the reductionist belief that all problems of organic form might be answered by dissecting organisms into separate features ..."[12] By the same token, no experienced linguist would ever claim that the order in which Latin, French or German words entered the English language explains very much, if anything, about its fundamental structure.

When cladistic analysis is applied to Cambrian paleontology, the imponderables of the method reappear as obscurities in the result, an interesting example of descent with modification. In this regard, Matzke writes, "the arthropods are instructive." And so they are. Let us be instructed. Let us be instructed by Gregory D. Edgecombe of the Department of Paleontology at the Natural History Museum in London. "Arthropod phylogeny," he writes cheerfully, "is sometimes presented as an almost hopeless puzzle wherein all possible competing hypotheses have support." His conclusions hardly amount to a ringing rejection of the hopeless puzzle school. The best that he can say for his field is that it is false that "anything goes."[13] I am sure this is so.

Nick Matzke is not about to go all hopeless on his supporters either. Witness his discussion of the otherwise hideous *Anomalocaris*. "Anyone actually mildly familiar with modern cladistic work on arthropods and their relatives," Matzke writes, "would realize that *Anomalocaris* falls many branches and many character steps below the arthropod crown group." As it happens, *Anomalocaris* does not fall anywhere: It is the anomalocaridids that do the falling. They in turn are folded within Radiodonta, which makes for an order and comprises a stem-group arthropod. Their most evident character in common with the arthropod crown group is what in a lobster would be called a bony claw. It is not much, but cladists are not fussy. The anomalocaridids include the genera *Anomalocaris, Peytoia, Schinderhannes, Amplectobelua* and *Hurdia*; the arthropod stem group, the gilled lobopodians, dinocaridids, the taxon incorporating Radiodontia, fuxianhuiids and canadaspidids. It is here that characters drift between Onychophora and the arthropod crown group itself.

Without ever mentioning just which shrimp he has mind, Matzke writes that "it is one of many fossils with *transitional* morphology *between* the crown-group arthropod phylum, and the next closest living crown group, Onychophora (velvet worms)." With this remark, he solidifies his reputation as a man capable of making the same mistake twice. Common characters? Or at least one in that bony claw? Yes, of course. *Transitional* morphology? *Ah,* but no. At best, an intermediate morphology. Nor can *it* be *intermediate* between the crown arthropods and Onychophora. That would be rather like placing a carrot as an intermediate between the United States and Bolivia. Wrong classification. It is Radiodonta as a taxon that is intermediate between taxa.

These are terminological disputes among us experts. A bloop is not necessarily a blunder. Let me refer in what follows to *Anomalocaris* X, where X designates whatever it is that Matzke had in mind. Does *Anomalocaris* X enter the fossil record after the first representatives of the arthropod crown make their appearance? It is in that case a little late, one might think, to be a transitional form. *Anomalocaris* X could hardly be ancestral to itself nor ancestral to the trilobites and other crown group arthropods. Before, then? It is in that case a little too complex to be the ancestor of the trilobites, possessing as do all such vile bugs compound eyes more sophisticated than anything exhibited by the trilobites— more sophisticated than anything except the eyes of various dragonflies, in fact. What, then, is the ancestor of *Anomalocaris* X? This is just the question that Stephen Meyer asks, again and again, as it happens. It is a part of the Cambrian mystery.

It is with a question such as this that the cladistic method achieves a triumph uniquely its own. We may allow Edgecombe the last word. It is in his Figure 1 that he displays a cladogram for stem and crown group arthropoda. The figure includes Onychophora, which falls outside these stem groups but is nonetheless hoping for cladistic glory. Thick black lines move downward from various stem groups and then they stop abruptly where the evidence leaves off. The cladogram nonetheless continues recklessly down through the muck and mist of the early Cambrian, thick black lines now replaced by lines that are thin. These mark the ghost lineages of the cladist's art, the artifacts of his method and not the imperatives of the evidence. While the last common ancestor of Radiodonta is basal to the last common ancestor of the arthropod crown, *both* are imaginary.

Ghost lineages are often defended, rarely extolled. Like much in cladistic analysis, they represent the withdrawal of a theory from any very robust confrontation with the evidence. They simply cannot be used to defend a view of the Cambrian that begins by questioning whether there is anything behind these ghosts beyond the cladist.

A man who believes in ghost lineages is demonstrably inclined, after all, to believe in ghosts.

Notes

1. Günter P. Wagner, *The Character Concept in Evolutionary Biology* (San Diego: Academic Press, 2001), xv.

2. Nick Matzke, "Meyer's Hopeless Monster, Part II," *Panda's Thumb*, June 10, 2013, http://pandasthumb.org/archives/2013/06/meyers-hopeless-2.html.

3. See Keynyn Brysse, "From weird wonders to stem lineages: the second reclassification of the Burgess Shale fauna," *Studies in History and Philosophy of Science Part C: Studies in History and Philosophy of Biological and Biomedical Sciences* 39, no. 3 (September 2008): 298–313; David A. Legg et al., "Cambrian bivalved arthropod reveals origin of arthrodization," *Proceedings of the Royal Society, Series B: Biological Sciences* 279, no. 1748 (2012): 4699–4704.

4. Willi Hennig, *Grundzüge einer Theorie der phylogenetischen Systematik* (Berlin: Deutscher Zentralverlag, 1950). No real cladist, it goes without saying, would ever refer to the English translation of this book. Let me see: I notice that Matzke never once refers to the German original. Readers must draw their own conclusions.

5. Make-work for jerks, as a distinguished figure was heard uncharitably to remark.

6. Andrew V. Z. Brower, "Evolution is not a necessary assumption of cladistics," *Cladistics*, 16, no. 1 (February 2000): 143–154. Later in his paper, Brower remarks that "systematics provides evidence that allows *inference* of a scientific theory of evolution." What doesn't?

7. See G. P. Wagner and P. F. Stadler, "Quasi-independence, homology and the unity of type: a topological theory of characters," *Journal of Theoretical Biology* 220, no. 4 (February 21, 2004): 505–527.

8. A point made vividly by Matzke's own source, which he cites in solemn incomprehension: Whatever the character matrix, Brysse observes, "there is only

enough reliable information available to construct cladograms, not trees." Brysse "From weird wonders to stem lineages."

9. M. Schmitt, "Claims and Limits of Phylogenetic Systematics," *Z. zool. Syst. Evolu.-forsch.* 27 (1989): 181–190. Schmitt is curator of Coleoptera and Head of Department of Arthropoda at the Zoologisches Forschungsmuseum Alexander Koenig. What he does not know about beetles is apparently not worth knowing.

10. "It is easy to calculate statistics such as CI and RI," Matzke writes, "and compare them to CI and RI statistics calculated based on data reshuffled under a null hypothesis where any possible phylogenetic signal has been obliterated." True enough. It is easy. "In virtually any real case, one will see substantial phylogenetic signal, even if there is uncertainty in certain portions of the tree." True enough again. The question of what the signal is signaling remains.

11. Matzke, "Meyer's Hopeless Monster Part II." By "bodyplan" Matzke means bodyplan. His remarks as written suggest someone about to dissect a word.

12. Stephen Jay Gould, *The Structure of Evolutionary Theory* (Cambridge: Harvard Univ. Press, 2002), 1057.

13. Gregory D. Edgecombe, "Arthropod Phylogeny: An Overview from the Perspective of Morphology, Molecular Data, and the fossil record," *Arthropod Structure & Development* 39 (2010): 74–87. This is not a paper that lends itself to Twitter. Edgecombe is not alone. The characters Matzke thinks homologous, Legg calls into question. See the caption, Legg. Legg is Matzke's witness and not ours. Valentine and Erwin do as much. Another witness, this one expert: "the lobopodians all share fairly simple, unspecialized legs, yet *Opabinia* and anomalocaridids lack legs but have paired, lateral flaps that, particularly in *Opabinia*, have gills along the upper aspect of the flap. Beyond the Radiodonta, however, well-sclerotized jointed appendages reappear. Are arthropod appendages homologous to those of lobopods, as Budd has argued? Are they homologous to the lateral flaps of Radiodonta [the group that includes anomalocaridids]? Or are they entirely novel structures? This debate is far from settled, illustrating the complexities of understanding the evolutionary pathways among these groups." Douglas Erwin and James Valentine, *The Cambrian Explosion: The Construction of Animal Biodiversity* (Greenwood Village, CO: Roberts and Company, 2013), 195. Internal citations removed.

8.

HOPELESS MATZKE

David Berlinski & Tyler Hampton

STEPHEN MEYER'S BOOK *DARWIN'S DOUBT* MAKES THREE CLAIMS: That the Cambrian explosion was real; that it remains unexplained; and that these facts sanction, or support, an inference to intelligent design. Writing at *Panda's Thumb*, Nick Matzke has denied the first, rejected the second, and ignored the third in "Meyer's Hopeless Monster, Part II." What a man ignores is his business. There is little point in demanding that Matzke assess an inference to which he is indifferent. If the Cambrian explosion was not real, Matzke is surely right to reject claims for its explanation. The Cambrian explosion, Matzke is persuaded, was a tedious, long-drawn Darwinian affair, one lasting for thirty million years and taking forever. This position has been warmly endorsed by Donald Prothero, an encomium, one might think, as impressive as one wrung from the lips of a Kardashian. Yet even if the Cambrian explosion was not real as an explosion, it remains real as an event. Cambrian creatures are in their body plan, nature, way of life, and order of complexity unlike their predecessors in the Ediacaran ooze. Having seen nothing new in the Cambrian era, it is hardly surprising that Matzke is persuaded that there is nothing new to see. "[T]here's no evidence," he writes, "that new protein domains were required in the Cambrian—I'd be surprised if any protein domains are known that are both unique to and required for the existence of Animalia."[1]

Writing in the *Proceedings of the National Academy of Sciences*, Susumo Ohno came to a different conclusion. He is disposed to see more
than Matzke, perhaps because his threshold of astonishment is lower. It
could not, of course, be higher. "It now appears," Ohno writes, "that this
Cambrian explosion, during which nearly all the extant animal phyla
have emerged, was of an astonishingly short duration, lasting only 6–10
million years." Thereafter Ohno draws the conclusion drawn in *Darwin's
Doubt*. New proteins did not originate in the Cambrian by means of an
incremental nibble. "[I]t is more likely that all the animals involved in
the Cambrian explosion were endowed with nearly the identical genome,
with enormous morphological diversities displayed by multitudes of animal phyla being due to differential usages of the identical set of genes."
What Ohno may have meant by the curious and suggestive word *endowed*, he does not say.[2]

The Cambrian genome was distinctive, Ohno argues, in five respects: It contained (i) a gene for lysyl oxidase, a protein that in the presence of molecular oxygen crosslinked collagen triple helices to produce
ligaments and tendons; (ii) genes for hemoglobin; (iii) genes for glass (si-
licified) skeletons; (iv) the Pax-6 gene for eye formation; and (v) a series of
Hox genes for the anterior-posterior (cranio-caudal) body plan. Ohno's
argument is offered on the level of molecular genetics, and *not* protein
chemistry, but if Cambrian animals required a gene for lysyl oxidase, it
is because they required lysyl oxidase as a protein. By parity of reasoning, if they incorporated a new and distinctive genome, presumably they
required a new and distinctive suite of proteins. The idea of a purely *deco-
rative* Cambrian genome is not a contribution to biology.

In an article published in *Science*, Chothia et al. added circumstances to what is largely a circumstantial case.[3] While 429 families of protein
domains are common to all eukaryotes, they argue, 136 are unique to
animals. These proteins presumably arose after the last common animal
ancestor. How otherwise could they be unique? If they arose after the
last common animal ancestor made his last stand, the victim, no doubt,
of passive smoking, they must have arisen throughout the Cambrian era.

By what mechanism of arousal? Chothia et al. offer only the vaguest of speculations. The evidence is suggestive; it is not conclusive. Yet suggestive evidence is better than none at all. Beyond appealing to surprise as a factor in his deliberations, Matzke has offered *no* evidence at all. He is in this regard serene.

In arguing against a Darwinian explanation for the emergence of complex new structures, whether in the Cambrian era or any other, *Darwin's Doubt* appeals to arguments of long-standing and continued controversy. Skeptics about Darwin's theory have for almost all of the twentieth century appealed to the same triplet to express their skepticism: the complexity of biological structures, the random nature of the search required to find them, and the limited time available to conduct the search.

Popular accounts of Darwinian theory have been devoted to mountain-climbing metaphors or mummeries, the progression by cumulative selection toward some cleanly defined optima.[4] Discussions have been pre-theoretical; indeed, they have been pre-scientific. They have barely counted as discussions.[5] No one need argue that complexity could be better defined, for complexity has never properly been defined at all. But for more than thirty years, it has become clear enough that hill-climbing is an irrelevance in evolutionary thought. The exquisite complexity of various biological structures cannot be explained by means of a strategy of no greater intellectual depth than Twenty Questions.[6] The path to virtually any complex structure must proceed by some scheme of deferred gratification.

In this regard, Matzke is of the Old School; his allegiances are to the Old Breed, biologists determined to face the facts by vigorously ignoring them. "The multiple-required-mutations stuff," he writes, "is basically just Behe's refuted 'irreducible complexity' argument disguised as an argument about sequence evolution, and is only relevant if it can be shown that 2 or more neutral mutations ever were required for anything relevant to the Cambrian Explosion, but, as is typical in DI literature, this is just blithely assumed rather than argued for."[7]

Michael Lynch is the last man on Earth likely to offer support to any theory of intelligent design. He is about as blithe as a canine incisor. His support is not needed. It quite suffices that he denies what Matzke affirms. "[A] broad subset of adaptations," he writes:

> cannot be accommodated by the sequential model, most notably those in which *multiple mutations must be acquired to confer a benefit*. Such traits, here referred to as complex adaptations, include the *origin of new protein functions involving multiresidue interactions*, the *emergence* of *multimeric enzymes*, the *assembly of molecular machines*, the colonization and refinement of introns, and the establishment of interactions between transcription factors and their binding sites, etc. The routes by which such evolutionary novelties can be procured include sojourns through one or more deleterious intermediate states.[8]

Is this "basically just Behe's argument disguised as an argument about sequence evolution?" It is. But with the caveat, of course, that Behe's argument, if it has been widely rejected by the Old Breed, has not been persuasively refuted by any of them—or by anyone else. "It is indeed true," Jerry Coyne remarked in *The New Republic*, "that natural selection cannot build any feature in which intermediate steps do not confer a net benefit on the organism."[9] Having recognized a challenge to right thinking in principle, Coyne is, of course, prepared to deny its significance in fact. Irreducible complexity? There is no such thing, Coyne remarks, and he has looked, too. In this, he is very much like a man wandering in a bakery and wondering petulantly where the bread might be. *Under your nose* is in both cases the instructive answer. Definitions are one thing; the real world, another; and irreducible complexity is not in doubt as a fact of life. A number of biologists like Michael Lynch quite understand this; they have pitched their tents in their enemies' camp. They appear to be quite at home.

Whether the Cambrian explosion was ignited by a short fuse or one that was long, the fact remains that having sputtered, the fuse in the end fired, the ensuing explosion producing new *Bauplans* or *Baupläne*, new phyla, new creatures entirely, such as *Hallucigenia*, all bristling porcupine quills and drooping snout, new skeletons, new tissues, and new nerves,

new brains and new eyes, the first to control the second and the second
to inform the first, and to accommodate all this newness, new genes and
the proteins they express, the staff and the stuff of life on every man's
account. Old Breeders see this as a part of the flow, the endless incom-
ing and outgoing tide of variation and incremental change, one protein
folding in onto itself and giving way to another. This thesis, *Darwin's
Doubt* rejects. The argument that it makes depends upon and amplifies
research conducted by Doug Axe and Ann Gauger.[10] Their conclusions
are similar to those reached by Mike Behe in *The Edge of Evolution*, but
careful historians of biology will notice that these arguments, when tak-
en collectively, have a distinctive history of their own, similar arguments
having been made long ago and then forgotten.[11]

Axe and Gauger begin by drawing a distinction to which critics such
as Nick Matzke are demonstrably insensible. The distinction is an essential one. "Functional innovations throughout the history of enzymes,"
Axe and Gauger write, "may be divided into two categories based on
the degree to which they depend upon structural innovation." Big Time
innovations are those contingent on a "fundamentally new structure"
and thus upon a new protein fold. Small Time innovations are matters
involving trimming or tightening or tinkering with an existing fold. It is
the difference between cutting a new pattern and embroidering an old
one. Axe and Gauger examined two proteins: KBl_2 and $BioF_2$. Although
structurally similar, they perform different functions. They are in their
active identities distinct. How difficult would it be to change one protein
so that it acquires the function of the other? To the extent that KBl_2
and $BioF_2$ are generic proteins, off the shelf and so off the cuff, the answer demanded by Darwinian theory is unequivocal: It should be pretty
easy. Darwin's theory of evolution is above any other consideration a
theory of *continuous* transformation and if small steps can do wonders
over the course of evolution, why not in the laboratory? One good question deserves another. How difficult would it be selectively to breed a cat
that barks in the night? The inescapable answer to both questions in the
world in which facts are plain is that it would be exceptionally difficult.[12]
Difficult enough, indeed, to prove unlikely in nature and impossible in

the laboratory. No matter the extent to which they tugged or pulled at KBl_2, Axe and Gauger never achieved a variant capable of executing the functions of $BioF_2$, the original protein remaining obdurate in its attachment to itself. The observation that whatever the selective pressure, a cat remains a cat, is as pertinent in protein chemistry as in zoology. If they were never able to do what they proposed, they *were* able to judge what it would require. A "successful functional conversion," they write, "would in this case require seven or more nucleotide substitutions."

They drew the obvious conclusion: The functional conversion of proteins is not possible under orthodox Darwinian scenarios. The job in question is too difficult for the time at hand.

Having read with irritation, or studied with indifference, Axe and Gauger's work, Matzke has remained adamant in his animadversions. If new proteins were needed in the Cambrian era, they were easily derived. Cambrian organisms had only to look around. Things were much easier than *Darwin's Doubt* suggests. Stephen Meyer has made far too much of far too little. *Ah*, the Old Breed. In this case, Matzke is not alone. He worships among a crowd of Old Believers, Jerry Coyne among them, genuflecting spastically. Axe and Gauger, they are persuaded, have surrendered to a pointless pessimism about what evolution might accomplish. If belief engenders belief, the facts are entirely less forthcoming. In 2003, Long et al. published an interesting paper entitled "The origin of new genes."[13] It is this paper to which Matzke appeals, often with satisfaction, always with assurance. Long et al. argued that the genes *Sdic*, *Sphinx*, *Jingwei*, *RNASE1*, and *AFGP*, among others, have originated, or evolved, within the last few million years, and this by what is essentially the old fond familiar process of typographic variation within the genome: Cut, Snip, Fiddle, Transpose, Shuffle, Replace. It is easy to see why Old Believers should find this paper rewarding. *It looks so simple.*

Those unforthcoming facts: In every example cited by Long et al., nothing like a new protein structure is in evidence; Long et al. remain throughout within the ambit of Small Time innovations, and from this ambit, they never depart, as the following discussion shows.

1) Sdic

A recently evolved chimeric gene encoding a protein in the species *Drosophila melanogaster*. Considering the order and sequence similarity of the surrounding genes, *AnnX* and *Cdic*, it is both easy and invigorating to imagine a process of duplication, deletion, fusion, and sequence rearrangements transforming *AnnX* and *Cdic* into *Sdic*.[14] Matzke says as much; he has lovely colored pictures to pass around. However, his analysis has been corrupted throughout by a systematic equivocation between genetic novelties and novelties in protein chemistry. The first is no good guide to the second. *Cdic* is a gene encoding a dynein intermediate polypeptide chain, one expressed in the cytoplasm. But *Sdic* *also* encodes a dynein-specific intermediate folding chain, one expressed in the testis. Save for the fact that it is truncated at the N-terminal, *Sdic* has the same structure as *Cdic*. It is expressed in sperm because the gene has acquired a new promoter from its fusion with *AnnX*.

The argument at issue concerns enzyme specificity and novel protein folds, *not* promoters and expression patterns.

In a stimulating paper entitled "Functional evidence that a recently evolved *Drosophila* sperm-specific gene boosts sperm competition," Shu-Dan Yeh et al. reported on work in which they deleted the region of the chromosome containing *Sdic* copies in order to test the effect of *Sdic* on fitness.[15] They subjected male flies with and without *Sdic* to mating rounds with females. The results were anything but provocative, least of all to the flies, and not at all to Shu-Dan Yeh et al. There was *no* discernible effect on the fly phenotype. Under benign conditions, those individuals without *Sdic* left as many progeny as those with *Sdic*. In multiple rounds of mating, when a female was first subjected to a mutant male without *Sdic*, and then one with, the presence of the gene gave no statistically significant advantage in sperm displacement. Repeating the experiment in reverse did lead to an advantage, but the relevance of this result to any larger issue of principle remained obscure. This is where it remains today.

The emergence of *Sdic* thus suggests a weak form of microevolution, so weak as to be embraced with equanimity by those proposing to champion, and those prepared to deny, the power of Darwinian theory. A protein dispensable to function, uninteresting in its chemistry, and with no control over morphology, has evolved by stochastic means. Whoever thought to deny it?

The lysyl oxidase family is otherwise.

Big difference.

2) Sphinx

A chimeric gene in *Drosophila* as well. Something of a hodgepodge, it was apparently scavenged from neighboring gene segments, a retroposed sequence of the ATP synthase F-chain gene from chromosome 2 obligingly inserting itself into the 102F region of chromosome 4. And the result? Nothing at all, as it happened. The issue is the subject of research conducted by Hongzheng Dai et al.[16] Their analysis followed the characteristic Darwinian trajectory in which, having prominently puffed up a claim, its authors were then obliged unobtrusively to puff it down. The puffing up: "[C]himeric genes often evolve rapidly," they write, "suggesting that they undergo adaptive evolution and may therefore be involved in *novel* phenotypes" (emphasis added). The puffing down: "[A]lthough [*Sphinx*] is derived, in part, from a protein-coding gene, it is most likely a noncoding RNA (ncRNA) because its parental-inherited coding regions are disrupted by several nonsense mutations."

A gene too impotent to produce a protein is hardly what Matzke requires to rebut *Darwin's Doubt*.[17]

He must do better.

He could not do worse.

3) Jingwei

Well-known for being well-known, *Jingwei* is a chimeric gene made from the Alcohol Dehydrogenase (*Adh*) gene and another called *yande*. The protein expressed by *Jingwei* is a member of the broad and noble class of enzymes that degrade alcohol (friends to humanity, if nothing else). Its parent gene, *Adh*, serves the same function. In a paper to which Long is a contributor, Jiaming Zhang et al. remark that

> *Drosophila Adh* belongs to the short-chain dehydrogenase/reductase (SDR) family ... SDRs share a common protein fold, consisting of a central β-sheet surrounded by α-helices and a typical nicotinamide coenzyme binding βαβαβ subdomain, with a characteristic Gly-Xaa-Gly-Xaa-Xaa-Gly motif ... Asp-37 confers specificity toward NAD binding, whereas the active site is characterized by a Ser-Tyr-Lys catalytic triad .... These and other conserved SDR features are preserved in *JGW*.[18]

The familiar deflationary wheeze now follows: "We predict that JGW will retain NAD-specific dehydrogenase activity."

No new folds. No new structures. No new nothing.

But, of course, something slightly different. When compared to *Adh*, it is plain enough that *Jingwei* is both more effective and more specific in degrading long-chain alcohols. These modest improvements in specificity did not simply emerge all at once, like Venus emerging from a clamshell in Botticelli's famous painting: *Adh* encodes a polypeptide that can metabolize a diverse range of alcohols as well. *Jingwei* thus *augments* the functions of an ancestral *Adh* enzyme. "Substrate specificity of *JGW*," Jiaming Zhang et al. write, "was further characterized in a survey of 34 alcohols that included representatives from all major classes found in nature... Like *Adh*, *JGW* shows activities toward a broad range of alcohols. However, compared with *Adh*, *JGW* also shows a systematic preference for long-chain primary alcohols and increased specificity... including farnesol and geraniol... These results confirm that *JGW* has evolved altered specificity after diverging from its parental genes, *Adh* and *ynd*."[19]

Improvements? Yes. Structural innovations? No. Relevance? None.

4) RNASE1

A gene destined for the short-term glory peculiar to genetic publicity campaigns. "The origination of new genes," Zheng et al. write, "was previously thought to be a rare event at the level of the genome ... However, it does not take many sequence changes to *evolve a new function.* [W]ith only 3% sequence changes from its paralogues, *RNASE1B* has developed a new optimal pH that is *essential* for the newly evolved digestive function in the leaf-eating monkey" (emphasis added).

This is interesting, exciting, and false.

a) The new and improved optimal pH promoted by *RNASE1B* does *not* involve a new function, let alone a new protein fold. The fold remains the Ribonuclease fold. Colobines are old world monkeys that eat leaves and employ symbiotic bacteria in their foregut to digest cellulose. The bacteria are themselves digested in the small intestine. To efficiently recycle nitrogen from RNA in quickly growing intestinal bacteria, it is better that the expression levels of *RNASE1B* be higher in the small intestine than in the foregut, and better that the optimal pH for its enzyme be lower, since the intestinal environment itself has a pH of between 6 and 7. The adaptation involved in lowering the optimal pH from 7.4 to 6.7 is accomplished principally by three forward mutations, something Zheng et al. demonstrated by reconstructing and then expressing the ancestral sequence in bacteria. Do the mutations altering *RNASE* confer a tangible benefit? They do not. There is *no* increased catalytic activity for the enzymes operating at the lowered optimal pH as opposed to the raised optimal pH. The mutations that lowered the optimal pH caused the enzyme to lose other features. Nothing in the account of Zheng et al. suggests that the mutations represented a complex adaptation.

b) Long's assertion that lowered optimal pH was *essential* for digestive function is a more urgent claim. It suggests the existence of tight functional constraints on changes leading to a complex adaptation. But whatever Long may suggest, the facts are otherwise. They so often are. "Although unproven," Zheng et al. write, "it is generally believed that foregut fermentation and leaf-eating emerged in the common ancestor of all colobines. Fossil evidence suggests that these changes occurred at least 10 Myr ago predating the duplications of *RNASE1*."

What follows is emphatic: "*The shift in optimal pH of the pancreatic RNases was not necessary for the changes in diet and digestive physiology of colobines.*"

Not necessary, meaning not essential. "Rather, the latter changes provided a selective pressure for more efficient digestive *RNases* in acidified environments, while gene duplication offered raw genetic materials that enabled this functional improvement."

Only the monkeys are apt to remain impressed.

5) AFGP

A member of the class of antifreeze glycoproteins used by Antarctic fish. Mike Behe is correct to note of these proteins that they have nothing to do with the demands of complex adaptation. Antifreeze proteins are not specific; they contain no robust secondary structure; they are poor in information and highly repetitive, surviving on a simple Thr-Ala-Ala repeat; and they do not interact with other proteins.

The features that *AFGP* employs to bind ice crystals would not work in forming lysyl oxidase. They would not work for any similarly specific protein family invented in the Cambrian. *That* required fold stability and well-defined tertiary structure for atomic-grade precision in orienting atoms to increase catalysis. *AFGP* is little more than a grunt-level blunt instrument.

Like so many other biologists, Matzke is persuaded that if in the modern protein theater, functional conversion is difficult, then in the ancient theater of the proteins, it must have been easy. This is as close to a transcendental deduction as an empirical science affords. But however valuable it may be as a metaphor, Deep Time is in the real world fickle as a factor, and while things may well have been different long ago, it hardly follows that they were *easier*. Various hypothetical scenarios tend to cancel one another. The play of forces, and the ensuing annihilation of advantage, is evident in Ohno's model of duplication and divergence. Where previously it had one gene, duplication provides an organism with two. One gene does the heavy cellular lifting and takes the obvious selective risks; the other is free to explore sequence space and serendipitously find new things to do. Genetic affairs do not get more flexible than this. This is surely a step in the right direction, no? It is by no means clear. Protein perturbation studies indicate that ~40% of all mutations "reduce or completely abolish function," a substantial portion (8%) leading to the "loss of *all* functions" (emphasis added). The rate of beneficial mutations, by way of contrast, stands at 103 or 0.1%. Absent selection, any duplicate will be crippled far faster than it will accumulate beneficial mutations.[20] Selection is, of course, unavailing. It is the *other* gene that is busy testing its luck in the real world. The ancient theater of the proteins may well have contained proteins flexible enough to *begin* things and cause a commotion, but what good a starting point if there is nowhere to go?

These considerations are hardly new. They have been long in rattling around the literature. The accommodating story is now of a sort standard in Darwinian theory in which a happy ending is demanded well before the story ends. "Evolutionary biologists agree," Austin L. Hughes remarks, "that gene duplication has played an important role in the history of life on Earth, providing a supply of novel genes that make it possible for organisms to adapt to new environments." Something makes it possible for organisms to adapt. This is the happy ending. "But it is less certain," he goes on to add, "how this panoply of new functions actually arises, leaving room for ingenious speculation but not much rigor." Ingenious speculations? Not much rigor? This is less happy. "Cases where we can reconstruct with any confidence the evolutionary steps involved in the functional diversification," he goes on to say, "are relatively few." Not much confidence? This is not happy at all.[21]

Ingenious speculations are required when it comes to protein evolution, because nothing better is available. Ancient proteins may well have been promiscuous in their affinities; mutations may have come in compensatory pairs, the good cancelling out the bad and vice versa; and those ancient proteins might well have embarked on weirdly successful random walks along neutral evolutionary networks. These phenomena are real enough, but they are more efficient *within* the context of a cell than they might otherwise be in *establishing* the context of a cell. A world in which fitness costs are steep is a world carrying ancestral organisms into fitness bankruptcy. How did these organisms survive? How did they do what organisms must do—condense chromosomes, translate proteins, harvest electrons? And to what constraining evolutionary pressures did they yield or succumb? We do not know. The questions are open.

What we *do* know is that questions of this kind are themselves promiscuous and reproduce freely. A recent study in *Science* indicates that in the case of antibiotic resistance, adaptive challenges can block a majority of Darwinian pathways. One relevant study showed that resistance to cefotaxime required 5 mutations. There are thus 5! =120 different trajectories for evolution to consider. But as the authors note: "In principle, evolution to this high-resistance beta-lactamase might follow any of the 120 mutational trajectories linking these alleles. However, we demonstrate that 102 trajectories are inaccessible to Darwinian selection and that many of the remaining trajectories have negligible probabilities of realization, because four of these five mutations fail to increase drug resistance in some combinations. Pervasive biophysical pleiotropy within the beta-lactamase seems to be responsible... we conclude that *much protein evolution will be similarly constrained*" (emphasis added).[22]

The prohibitive force is sign epistasis, an extreme and open-ended form of context dependence. "Sign epistasis means," write the authors who coined the term, "that the sign of the fitness effect of a mutation is under epistatic control; thus, such a mutation is beneficial on some genetic backgrounds and deleterious on others."[23] If evolution is so easily confounded by an antibiotic challenge, how much more stringently will it be constrained given the intuitively harder, multi-level challenges involved in creating the histone protein complex? Or various Cambrian organisms, for that matter? In the absence of relevant epistatic factors, who knows? Were evolutionary biologists not professionally committed to happy endings, they would acknowledge with grace that the modern protein theater emerged by means of conditions that we cannot specify in organisms whose nature we cannot imagine.

This is, after all, what Axe, Gauger, and Stephen Meyer are saying. It is not all that they are saying; but *à chaque jour suffit sa peine*.

Notes

1. Nick Matzke, "Meyer's Hopeless Monster, Part II," *Panda's Thumb*, June 10, 2013, http://pandasthumb.org/archives/2013/06/meyers-hopeless-2.html.

2. Susumo Ohno, "The notion of a pananimalia genome," *Proceedings of the National Academy of Sciences, USA* 93, no. 16 (August 6, 1996), 8475–8478. Ohno is well-known for his thesis that evolutionary change proceeds by gene duplication and divergence, what is now called the Ohno model. See Susumo Ohno, *Evolution by Gene Duplication* (Berlin: Springer-Verlag, 1970).

3. Cyrus Chothia et al., "Evolution of the Protein Repertoire," *Science*, 300, no. 5626 (June 13, 2003): 1701–03.

4. There is no completely general and mathematically rigorous account of Darwinian theory, a point ceded by mathematical biologists on those occasions when they are free to whisper into one another's ears. "Darwin's theory of evolution by natural selection has obstinately remained in words since 1859. Of course, there are many mathematical models that show natural selection at work, but they are all examples. None claims to capture Darwin's central argument in its entirety." A. Grafen, "The Formal Darwinism Project: A Midterm Report," *J Evol Biol*. 20, no. 4 (July 2007): 1243–54. The phrase 'obstinately remained in words' must be assigned its proper meaning: obstinately remaining on the level of anecdote, example, and gossip.

5. It is worth observing that Fred Hoyle anticipated Mike Behe's idea of an irreducibly complex system in his discussion of the Histone protein complex. See his *Mathematics of Evolution* (Memphis: Acorn Enterprises, 1999). Hoyle's derivation of the principles of population genetics is well worth reading, especially his discussion of Kimura's diffusion equations. His book is, needless to say, widely inaccessible.

6. Responding to skeptics at the Wistar Symposium such as Murray Eden and M. P. Schützenberger, Sewall Wright appealed to Twenty Questions as evidence they had overlooked a fast algorithm for the generation of complexity—not his finest moment.

7. Matzke, "Meyer's Hopeless Monster, Part II."

8. Michael Lynch, "Scaling expectations for the time to establishment of complex adaptations," *Proceedings of the National Academy of Sciences* 107, no. 38 (September 7, 2010): 16577–82. Emphasis added.

9. Jerry Coyne, "The Great Mutator," *The New Republic* (June 14, 2007), 42.

10. Doug Axe and Ann Gauger, "The Evolutionary Accessibility of New Enzyme Functions: A Case Study from the Biotin Pathway," *BIO-Complexity* 2011, no. 1 (April 11, 2011): 1–17, http://bio-complexity.org/ojs/index.php/main/article/view/BIO-C.2011.1.

11. Michael J. Behe, *The Edge of Evolution* (New York: Free Press, 2007). Long ago? See Fred Hoyle's *The Mathematics of Evolution* (1999).

12. It is not impossible to transform a base metal into gold—just very difficult. From alchemy to atomic theory is a progression governed in part by a sliding parameter, one measuring the difficulty of atomic transmutation. The ancients thought it easy, the moderns think it hard. In evolutionary biology, it is the other way around.

13. Manyuan Long et al., "The origin of new genes: glimpses from the young and old," *Nature Reviews Genetics* 4, no. 11 (November 2003): 865–875. *Darwin's Doubt* contains a fine discussion of Long et al., one starting on p. 222. Of course it does.

14. See Rita Ponce & Daniel L. Hartl, "The evolution of the novel Sdic gene cluster in *Drosophila melanogaster*," *Gene* 376, no. 2 (July 19, 2006): 174–183.

15. Shu-Dan Yeh et al., "Functional evidence that a recently evolved Drosophila sperm-specific gene boosts sperm competition," *Proceedings of the National Academy of Sciences, USA* 109, no. 6 (February 7, 2012): 2043–48.

16. Hongsheng Dai et al., "The evolution of courtship behaviors through the origination of a new gene in Drosophila," *Proceedings of the National Academy of Sciences, USA* 105, no. 21 (May 27, 2008): 7478–83.

17. That non-coding RNA may play any number of important roles in the cell is a separate issue, one no longer in doubt but equally of no relevance to the point under discussion.

18. Jiaming Zhang et al., "Evolving protein functional diversity in new genes of Drosophila," *Proceedings of the National Academy of Sciences, USA* 101, no. 46 (November 16, 2004): 16246–50.

19. Internal references to illustrations have been deleted.

20. Misha Soskine, M. and Daniel S. Tawfik, "Mutational effects and the evolution of new protein functions," *Nature Reviews Genetics* 11 (August 2010): 572–82. "Moreover, for a significant fraction of proteins, increased dosages result in reduced fitness owing to undesirable promiscuous interactions driven by high protein concentrations or disturbed balance of protein complexes. Thus, *although increased protein doses can make a weak, promiscuous activity come into action and thereby provide an evolutionary starting point, these increased doses may also become deleterious owing to the very same effect.*" (Emphasis added.) This might be called the Monkey's Paw effect in molecular genetics.

21. Austin L. Hughes, "Gene Duplication and the Origin of Novel Proteins," *Proceedings of the National Academy of Sciences, USA* 102, no. 25 (June 21, 2005): 8791–92. Hughes does mention an example of a happier ending: "Thus the report in this issue of *PNAS* by Tocchini-Valentini and colleagues on tRNA endonucleases of Archaea is particularly welcome as a concrete example of how new protein functions can arise." *Can arise*, note, not *has arisen*.

22. Daniel M. Weinreich et al., "Darwinian Evolution Can Follow Only Very Few Mutational Paths to Fitter Proteins," *Science* 312, no. 5770 (April 7, 2006): 111–14.

23. See Daniel M. Weinreich, R. A. Watson, and L. Chao, "Sign Epistasis and Genetic Constraints on Evolutionary Theory," *Evolution* 59, no. 6 (June 2005): 1165–74. Weinreich et al. would appear to be appealing to the concept of a context-sensitive grammar in order to account for, or describe, sign epistasis. A context-sensitive grammar is one whose production rules are of the form $\alpha A \beta \rightarrow \alpha \gamma \beta$, where the derivation of γ from A depends on the flanking parameters α and β.

9.

CLADISTICS TO THE RESCUE?

Casey Luskin

FOR THE PAPERBACK EDITION OF *DARWIN'S DOUBT*, STEPHEN MEYer added an Epilogue, answering the more substantive of his critics including Nick Matzke, Donald Prothero, and Charles Marshall. Matzke's original response to Meyer at *Panda's Thumb* argued that cladistics, a method of phylogenetic analysis that generates tree diagrams called cladograms, can show purported relationships among species, and dispel the problem posed by the absence of fossils ancestral to the animals that appear in the Cambrian explosion, demonstrating their stepwise, gradual evolution.

In the Epilogue, Meyer narrowly circumscribes his reply to Matzke. Since he was not trying to argue against common descent, he emphasizes that cladistics cannot compensate for the absence of ancestral fossils. Meyer thus makes four main arguments in response to Matzke:

First, cladistics cannot do what Matzke claims it can do—explain the absence of ancestors—since it presupposes the very thing that it attempts to prove. It assumes that there is a common ancestor and then seeks to prove the existence of a common ancestor. This begs the question.

Second, because of the sequence of appearance of crown and stem groups, cladistic analysis, if interpreted as an historical account of events, forces evolutionary biologists to draw lengthy lines of ancestry representing ghost lineages. As Meyer points out, rather than solving the problem of why fossils are missing, this creates a situation that requires *more* missing fossils.

Third, there are multiple competing histories of life that are compatible with any given cladogram. This means cladistics cannot, in principle, establish a particular evolutionary history, or specific ancestor-descendant relationships between organisms, as Matzke claims it can.

Fourth, at best, cladistic methodology identifies a pattern of relationships between organisms, or a branching pattern showing the order in which traits arose. But cladistic analysis does not and cannot explain the *causes* of the origin of those characters that it classifies. Therefore it cannot explain the origin of the Cambrian animals, or the information required to build them. As Meyer explains in the Epilogue:

> Cladistics describes patterns of relationships among organisms; it provides tools for classifying organisms. It might also suggest historical reconstructions of evolutionary history if its question-begging assumptions in that context are granted. But it cannot determine what caused the patterns of relationship depicted by cladograms or what *caused the origin of the complex animal features* that it analyzes. For this reason, cladistics cannot be used to rebut the central argument of *Darwin's Doubt*, which addresses precisely the question of what caused the Cambrian animals to arise.

And that is why Matzke's review of *Darwin's Doubt* fails to address the central argument of the book. Cladistics does not, and cannot, offer any explanation of what caused the Cambrian animals to come into existence. Nor can it account for the origin of genetic and epigenetic information necessary to produce them.[1]

Meyer observes that one of Matzke's primary authorities on cladistic thinking, Keynyn Brysse, makes these very points. "Cladistics," Brysse writes, "cannot be used to judge" the "tempo and mode of evolution" or "the nature of the evolutionary process."[2]

Matzke posted a response to Meyer's Epilogue, but it doesn't even attempt to address these arguments in any relevant detail. It is not an adequate rebuttal in any way, shape, or form. Specifically:

Regarding argument (1), Matzke attempts no rebuttal.

Regarding argument (2), Matzke's response is irrelevant. Rather than explaining why animal phyla appear abruptly in the Cambrian, he writes about the origin of mammals and dinosaurs—topics that aren't even discussed in *Darwin's Doubt*.

Regarding arguments (3) and (4), Matzke repeats his prior claims that there are transitional fossils showing evolutionary steps in the origin of Cambrian animals, especially arthropods. He thinks this branching pattern demonstrates that arthropod traits evolved in a gradual, step-wise manner, which in his mind shows that the information required to build new animal body plans arose via Darwinian mechanisms. Matzke writes:

> Meyer repeats his statements about how cladistics doesn't show how new information and developmental changes come about— No, cladistics shows the major steps that occurred and their order, and that disproves the idea that it had to happen all at once in defiance of Darwinian gradualism, which is a key feature of Meyer's argument.[3]

Yet Matzke fails to address Meyer's central points—and the authorities Meyer cites affirming them—that cladistic analysis cannot elucidate a specific evolutionary history or the causes of the patterns it generates. Matzke's argument commits the *post hoc ergo propter hoc* fallacy. He believes (wrongly) that a pattern (or correlation) establishes the cause of the pattern. However, correlation does *not* demonstrate causation.

Matzke's argument is thus fundamentally powerless to rebut the central argument of *Darwin's Doubt*. Cladistics cannot explain the causes of the pattern it finds. It also fails to elucidate a clear and consistent pattern of major evolutionary steps, especially with regard to Matzke's example of Cambrian arthropods, as we shall now see.

On cladistics, Meyer's main point is that the method cannot explain the causes or processes that generated the purported relationships between Cambrian arthropods that cladistic analysis identifies. But cladistics has also failed to demonstrate a branching pattern showing the ordering of the major steps of evolution in Cambrian arthropods.

For example, in earlier replies to *Darwin's Doubt*, Matzke claimed that *Anomalocaris* and other members of its family (the anomalocaridids) are "intermediate" to arthropods. Why? Because they share some (though not all) of the derived characteristics that define arthropods. He asserts that *Anomalocaris* "is one of many fossils with transitional morphology between the crown-group arthropod phylum, and the next closest living crown group, Onychophora (velvet worms)."[4] But leading authorities doubt that the characteristics supposedly linking anomalocaridids to arthropods are homologous—in other words, that they demonstrate shared ancestry.

For example, in his initial response to Meyer, Matzke posted a cladogram from David Legg et al. (2012) linking *Anomalocaris* to arthropods because of their similar compound eyes and cephalic limbs (limbs extending from their heads). But he failed to include the caption, which admits serious doubts about homology between these characters, noting that compound eyes "disappeared and 're-evolved' a number of times," and that "it is unclear if the arthropodized cephalic limbs of radiodontans (*Anomalocaris* and *Hurdia*) are homologous to the arthropodized trunk limbs of arthropods."[5] In another article from 2013, Gregory Edgecombe and Legg even suggest that anomalocaridids might be "non-arthropods" with only "convergent similarity to arthropods" since their "affinities to Arthropoda are controversial."[6]

Similar problems exist with regard to another supposed "transitional form" Matzke cites—the lobopods, represented today by velvet worms, of phylum Onychophora. He claims that "the arthropod and velvet-worm phyla evolved from lobopods, and lobopods contain a whole series of transitional forms showing the basics of how this happened." Moreover, he asserts that "*Anomalocaris*... is one of many fossils with transitional morphology *between* the crown-group arthropod phylum, and the next closest living crown group, Onychophora (velvet worms)."

This suggests to him that *Anomalocaris* was intermediate between lobopods and arthropods. Matzke further claims that "cladistic analyses reveal the order in which the characters found in living groups were acquired."[7] Yet lobopods and anomalocaridids are a prime example of how cladistics has failed to reveal such ordering.

If, according to Matzke, both lobopods and radiodontans (the order that includes anomalocaridids) are closely related to arthropods, which group is *more* closely related to arthropods? To put the question another way, what is the *order* in which these two groups branch off from the tree leading to arthropods? Experts disagree on this question, and their reasons for disagreeing challenge the homology of lobopod and anomalocaridid features with arthropods.

Arthropods are defined by a number of features. But understanding the origin of arthropods requires explaining how they acquired paired, jointed appendages (legs) and a complex head and compound eyes. Lobopods have legs, sometimes resembling paired and jointed appendages in arthropods, but they lack arthropod-like heads with compound eyes. Anomalocaridids lack legs entirely, but some had arthropod-like heads with compound eyes. Sharp disagreement exists over whether lobopods (with legs, not heads or compound eyes), or anomalocaridids (with heads and compound eyes, not legs), are more closely related to arthropods. Cladistic analyses have failed to yield a consensus on the branching "order" in which these crucial arthropod characters were acquired.

Roughly speaking, the situation is like Figure 9-1 below. Cladists must decide which is more closely related to arthropods—A or B?

If you answer "A," and the legless anomalocaridids are more closely related to arthropods, then this also means the legs of lobopods cannot be homologous to arthropod legs (and cladists have no ancestral source to explain the evolution of arthropod legs). If you answer "B," and the compound-eyeless lobopods are closer to arthropods, then the compound eyes of anomalocaridids cannot be homologous to arthropod eyes (and cladists have no ancestral source to explain the evolution of arthropod eyes). There's no good solution for cladists. Pick either option, and you're faced with a situation where one of Matzke's so-called "transitional" forms *cannot be* transitional, and may not even belong in the branching pattern of organisms leading to arthropods.

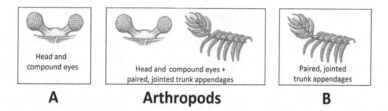

Figure 9-1. Illustration: Image created by Casey Luskin using components with permission from other papers. *Diania* image on right adapted from Macmillan Publishers Ltd. Jianni Liu, Michael Steiner, Jason A. Dunlop, Helmut Keupp, Degan Shu, Qiang Ou, Jian Han, Zhifei Zhang, & Xingliang Zhang, "An armoured Cambrian lobopodian from China with arthropod-like appendages," *Nature*, 470 (24 February 2011), 526-530, fig. 3. Adapted by permission of Macmillan. Anomalocaridid head on left adapted from David A. Legg, Mark D. Sutton, Gregory D. Edgecombe and Jean-Bernard Caron, "Cambrian bivalved arthropod reveals origin of arthrodization," *Proceedings of the Royal Society B*, 279 (2012), 4699–4704. Figure 3. By permission of the Royal Society.

Figure 9-2 further illustrates this dilemma.

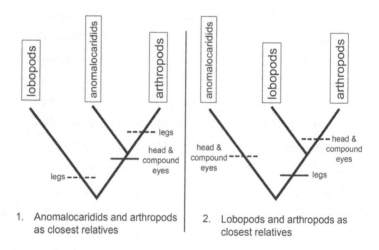

Figure 9-2. Illustration: Paul Nelson.

As seen in Figure 9-2, cladists must choose between option (1) and (2), but in either case a key arthropod trait lacks homology with another Cambrian animal group and must have evolved convergently. The solid horizontal lines indicate an hypothesis of genuine homologies; the dotted horizontal lines would then be homoplastic (convergent) characters.

Matzke's rebuttals to Meyer disclose none of these problems. However, a brief review of the literature shows these issues are widely recognized. A *Nature* article from 2011 notes that "potential stem-arthropods typically express **mosaics** of arthropod-like characters, which makes resolving a single, simple tree of arthropod origins problematic."[8] Another *Nature* paper frames the problem this way:

> [T]he absence of sclerotized trunk limbs in dinocaridids (*Anomalocaris*, etc.)... **remains for us one of the great puzzles of stem-group arthropod evolution.** Put simply, dinocaridids [the class which includes anomalocaridids] have a more arthropod-like head region (cephalisation, eyes, sclerotized mouthparts), whereas *Diania* [a lobopod] lacks such sophistication in the anterior body [head] region, but has jointed trunk appendages.[9]

Another paper in the same exchange directly contradicts Matzke's claim that cladistics has revealed the ordering of arthropod evolution:

> The increasingly detailed fossil record of stem-group euarthropods provides our best chance of resolving this issue, but as yet has failed to do so; **unequivocal evidence for any particular ordering of acquisition in these characters is not yet available.**[10]

Moreover, unique and highly derived features of anomalocaridids and lobopods make them unlikely candidates for arthropod precursors. *Anomalocaris* has weird flexible lobes on each side of its body, and a unique mouth with ringed teeth. As two experts explain, it "is unlike any known arthropod, particularly in the nature of the jaw apparatus and the close-spaced, strongly overlapping lateral lobes."[11] Lobopods have equivalent peculiarities. One paper discussing *Diania*, a lobopod that some have claimed is closely related to arthropods, observes that it is "a highly unusual creature. It is hard to envisage it as the progenitor of any modern arthropod group."[12] In constructing cladograms showing these organisms as intermediate, cladistic analysis finds it convenient simply to ignore such unique features.[13]

Authorities Would Agree: Matzke Overstates His Cladistics Case

Multiple authorities have recognized that Matzke's smooth evolutionary grade of intermediates leading to arthropods does not exist. Douglas Erwin and James Valentine's 2013 tome *The Cambrian Explosion*, which we refer to often in this book, summarizes the debate by observing that the homology of anomalocaridid or lobopodian appendages with arthropods is one of "the problems currently facing researchers." The write that "[t]his debate is far from settled, illustrating the complexities of understanding the evolutionary pathways among these groups."[14]

Likewise, Stefan Richter et al. (2013) note: "The exact composition of the stem group of arthropods depends on the position of the Onychophora and Tardigrada, which to date remains unresolved" and "there are still taxa which are the subject of debate with regard to their phylogenetic position, one being the 'great appendage arthropods'"—e.g., the anomalocaridids.[15]

In other words, the phylogenetic positions of both anomalocaridids and lobopods, as well as the ordering of the evolution of arthropod characters, are unestablished. Similarly, Gregory Edgecombe writes: "Arthropod phylogeny is sometimes presented as an almost hopeless puzzle wherein all possible competing hypotheses have support."[16] Even Matzke's citation from Legg et al. (2012) explains: "The origin of arthropods is a contentious issue... there is little consensus regarding the details of their origins."[17] Meyer cites many of these authorities in his Epilogue, yet none of Matzke's replies address or acknowledge these problems.

In earlier responses to Meyer, Matzke had claimed arthropods are "instructive" in showing "the major steps that occurred and their order" and that "cladistic analyses reveal the order in which the characters found in living groups were acquired." Arthropods are instructive, but not in the way Matzke meant it. The literature refutes Matzke's assertions that cladistics has shown the ordering of the evolution of arthropod traits.

Consistency Indices and Straw Men

As I noted, in his initial reply to *Darwin's Doubt*, Matzke posted cladograms showing the relationships between various stem and crown group arthropods. In the Epilogue, Meyer responds by pointing out that these cladograms had low consistency indices (CIs), a measure of how often the assumption of homology—that similarity results from common ancestry—fails within a tree. Low CIs (closer to zero) mean that biological similarity frequently is *not* the result of common ancestry, while higher CIs (closer to one) suggest the data fits a treelike pattern.

Meyer's point wasn't that low CIs necessarily refute common ancestry, but rather that they show how often the assumptions of cladistic analysis fail, undermining the strength of its conclusions. As Meyer observed, in the cladograms Matzke posted, the assumption of homology failed 43 percent and 61 percent of the time.[18] Even the authors who created one of the cladograms touted by Matzke noted their CI was "rather low."[19]

Again, Meyer's main purpose here isn't to claim that common descent is therefore wrong. Rather, his main argument is that cladistics cannot compensate for the absence of ancestral fossils. His second argument is that the assumptions of the cladistics methods frequently fail. If common descent is true, it isn't demonstrated by Matzke's cladograms.

In response to the "low" CIs of his cited cladograms, Matzke offered this retort: "So what?" He then sets up a straw man null hypothesis for testing common ancestry, claiming we can only question common ancestry when we find a *random* distribution of traits. Matzke cites a paper that measured the CIs of randomized datasets of various sizes. As long as the CI is higher than what's predicted by a random dataset of the same size as your dataset, Matzke argues, common ancestry is a sound conclusion.

Matzke confuses the issue. In the paper he cited, the null hypothesis isn't the refutation of common ancestry, but a *completely random dataset*. As G. J. Klassen et al. (1991) explain, "This study was undertaken to determine the range of consistency index (CI) values obtainable from

random data sets." But does a CI that's higher than what you'd expect from a random dataset necessarily demonstrate common ancestry? They caution against this conclusion:

> This is not to say that we believe any data sets exhibiting CIs greater than the marked confidence intervals are 'good.' The high random data CI values obtained for smaller data sets are a poignant reminder of the caution required when interpreting tree topologies.[20]

This is insightful: one might statistically reject a "random" dataset, yet still have a mediocre CI that is unhelpful in elucidating the true phylogenetic pattern.

By suggesting that we cannot reject common ancestry unless the CI is so low that it implies randomized data, Matzke erects an unreasonably high standard for questioning the conclusions of cladistic analysis. He claims "creationists make the perfect the enemy of the good," but he's saying we must tolerate the marginal until we find the absolutely horrible.

Critics would never say we can only refute common ancestry if the data is so bad that it's "random." Non-common-ancestry-based datasets could frequently have higher-than-random CIs.

For example, I blindly grabbed 12 shirts from my closet and scored them for 16 traits (e.g., buttons, zipper, solid color, etc.). While scoring the data, I immediately observed non-random correlations that would give my dataset a decent CI. For example, shirts with buttons are often associated with a solid color, a pocket, and a collar. My tree had a CI of 0.76—*much higher* than ~0.25, what G. J. Klassen et al. (1991) say we should expect from a random dataset of that size. Yet common ancestry did not generate the shirts in my closet! What does this show? A dataset with a higher-than-random CI doesn't necessarily imply common ancestry. Why? Common design predicts re-usage of parts in a non-random manner that fulfills design constraints required by the system.

After observing that the assumption of homology failed over 60 percent of the time in one of Matzke's cladograms, Meyer asks: "If an assumption fails more often than it holds true, is it justified?" That's a fair question, and Matzke's cavalier "So what?" confirms that cladistics is a field that tolerates large amounts of data that contradict its own assumptions.

Notes

1. Stephen C. Meyer, *Darwin's Doubt: The Explosive Origin of Animal Life and the Case for Intelligent Design* (New York: HarperOne, 2014). This is in the Epilogue added to the paperback edition, page 437.

2. Keynyn Brysse, "From weird wonders to stem lineages: the second reclassification of the Burgess Shale fauna," *Studies in History and Philosophy of Biological and Biomedical Sciences* 39, no. 3 (September 2008): 298–313.

3. Nick Matzke, "Meyer's Hopeless Monster, Part III," June 5, 2014, http://pandasthumb.org/archives/2014/06/meyers-hopeless-3.html.

4. Nick Matzke, "Meyer's Hopeless Monster, Part II," *Panda's Thumb*, June 19, 2013, http://pandasthumb.org/archives/2013/06/meyers-hopeless-2.html.

5. David A. Legg et al., "Cambrian bivalved arthropod reveals origin of arthrodization," *Proceedings of the Royal Society B* 279 (2012): 4699–4704.

6. Gregory Edgecombe and David Legg, "The Arthropod Fossil Record," in *Arthropod Biology and Evolution* (Berlin: Springer, 2013): 393–415.

7. Matzke, "Meyer's Hopeless Monster, Part II."

8. Jianni Liu et al., "An armoured Cambrian lobopodian from China with arthropod-like appendages," *Nature* 470 (February 24, 2011), 526–530. Emphasis added.

9. Liu et al., "Liu et al. reply," *Nature*, 476:e3-e-4 (August 11, 2011). Emphasis added.

10. Legg et al., "Lobopodian phylogeny reanalysed," *Nature*, 476:E2 (August 11, 2011). Emphasis added.

11. H. B. Whittington and D. E. G. Briggs, "The Largest Cambrian Animal, Anomalocaris, Burgess Shale, British Columbia," *Philosophical Transactions of the Royal Society B* 309, no. 1141 (May 14, 1985): 569–609.

12. Liu et al., "An armoured Cambrian lobopodian," 530.

13. See Derek E. Briggs and Richard A. Fortey, "The Early Radiation and Relationships of the Major Arthropod Groups," *Science* 246 (October 13: 1989): 241–243; K. Brysse, "From weird wonders to stem lineages: the second reclassification of the Burgess Shale fauna," *Stud. Hist. Philos. Biol. Biomed. Sci.* 39, no. 3 (September 2008): 298–313; Stephen Jay Gould, "The Disparity of the Burgess Shale Arthropod Fauna and the Limits of Cladistic Analysis: Why We Must Strive to Quantify Morphospace," *Paleobiology* 17, no. 4 (Autumn 1991): 411–423.

14. Douglas Erwin and James Valentine, *The Cambrian Explosion: The Construction of Animal Biodiversity* (Greenwood Village, CO: Roberts and Company, 2013), 195, 202. (Internal citations removed.)

15. Stefan Richter et al., "The Arthropod Head," in Alessandro Minelli et al., *Arthropod Biology and Evolution: Molecules, Development, Morphology* (Berlin: Springer, 2013), 223–240.

16. Gregory Edgecombe, "Arthropod Phylogeny: An Overview from the Perspective of Morphology, Molecular Data, and the Fossil Record," *Arthropod Structure & Development* 39 (2010): 74–87.

17. Legg et al., "Cambrian bivalved arthropod."

18. *Darwin's Doubt*, "Epilogue," 432–433 (paperback edition).

19. Briggs and Fortey, "The Early Radiation and Relationships," 242.

20. G. J. Klassen et al., "Consistency Indices and Random Data," *Systematic Zoology* 40, no. 4 (January 1991): 445, 455. Emphasis added.

III.

HEAVYWEIGHT: BERKELEY'S CHARLES MARSHALL

There is no credible scientific challenge to the theory of evolution as an explanation for the complexity and diversity of life on Earth.

NEW YORK TIMES
Cornelia Dean, "Evolution Opponent Is in Line for Schools Post," *New York Times*, May 19, 2007, http://www.nytimes.com/2007/05/19/education/19board.html.

10.

WHEN THEORY TRUMPS

OBSERVATION

Stephen C. Meyer

I AM PLEASED THAT THE JOURNAL *SCIENCE* ASSIGNED MY BOOK *DAR-win's Doubt* for review to the distinguished UC Berkeley paleontologist Charles Marshall. His is the first critical review ("When Prior Belief Trumps Scholarship") to grapple with the book's main arguments about the inability of standard evolutionary mechanisms to explain the origin of morphological novelty in the Cambrian period. Though Marshall does address the main problem discussed in the book, his review also demonstrates—if inadvertently—the severity of that problem, and that leading Cambrian paleontologists and evolutionary biologists (such as Dr. Marshall) are nowhere close to solving it.

In this and succeeding chapters, I will offer four responses to the four main critiques that Marshall presents of *Darwin's Doubt* in his review. I begin here by addressing Marshall's claim that developmental gene regulatory networks (dGRNs), which are necessary for the development of animals, could have been more labile or flexible in the past. Marshall made this claim to challenge my contention in Chapter 13 of *Darwin's Doubt* that the observed inflexibility of these regulatory networks represents a major impediment to the evolutionary transformation of one animal body plan into another. Not so, he argues. As he asserted in his review, "Today's GRNs have been overlain with half a billion years of evolutionary innovation (which accounts for their resistance to modification), whereas GRNs at the time of the emergence of the phyla were not so encumbered."[1]

Yet, contrary to Marshall's speculation about how dGRNs might have functioned in the past, all available observational evidence establishes that dGRNs do not tolerate random perturbations to their basic control logic. Indeed, mutagenesis experiments conducted on the genes present in dGRNs have repeatedly shown that even modest mutation-induced changes to these genes either produce no change in the developmental trajectory of animals (due to pre-programmed buffering or redundancy) or they produce catastrophic (most often, lethal) effects within developing animals. Disrupt the central control nodes, and the developing animal does not shift to a different, viable, stably heritable body plan. Rather, the system crashes, and the developing animal usually dies. As developmental biologist Eric Davidson has noted:

> There is always an observable consequence if a dGRN subcircuit is interrupted. Since these consequences are always catastrophically bad, flexibility is minimal, and since the subcircuits are all interconnected, the whole network partakes of the quality that there is only one way for things to work. And indeed the embryos of each species develop in only one way.[2]

Thus, to claim, as Marshall does in his review of *Darwin's Doubt*, that dGRNs might have been more elastic in the past contradicts what developmental biologists have learned over several decades of investigating how these networks actually function from mutagenesis studies of many different biological "model systems," including *Drosophila* (fruit flies), *Caenorhabditis* (nematodes), *Strongylocentrotus* (sea urchins), *Danio* (zebrafish) and other animals.[3] Moreover, as Marshall himself has noted elsewhere, there is a good reason for this inflexibility. As Marshall explains, "many of the characters that evolved during the origin of the phyla are no longer able to change. The reason for this is that selectable variation is absent: either *the characters are invariant* or *mutants that carry this variation are sterile or lethal.*"[4]

Of course, Marshall thinks that these networks could have had a different, more flexible, character in the past. Yet, given what dGRNs do, namely, enable different cell types to organize themselves, and differentiate themselves from each other, in precise ways at precise times during the development of specific animal forms, it is hard to see how dGRNs could have functioned as regulatory networks and also exhibited the kind of flexibility that Marshall envisions. Developmental gene regulatory networks are control systems. A labile dGRN would generate (uncontrolled) variable outputs, precisely the opposite of what a functional control system does. It is telling that although many evolutionary theorists (like Marshall) have speculated about early labile dGRNs, no one has ever described such a network in any functional detail—and for good reason. No developing animal that biologists have observed exhibits the kind of labile developmental gene regulatory network that the evolution of new body plans requires. Indeed, Eric Davidson, when discussing hypothetical labile dGRNs, acknowledges that we are speculating "where no modern dGRN provides a model" since they "must have differed in fundamental respects from those now being unraveled in our laboratories."[5]

By ignoring this evidence, Marshall and other defenders of evolutionary theory reverse the epistemological priority of the historical scientific method as pioneered by Charles Lyell, Charles Darwin, and others.[6] Rather than treating our present experimentally based knowledge as the key to evaluating the plausibility of theories about the past, Marshall uses an evolutionary assumption about what must have happened in the past (transmutation) to justify disregarding experimental observations of what does, and *does not*, occur in biological systems. The requirements of evolutionary doctrine thus trump our observations about how nature and living organisms actually behave. What we know best from observation takes a back seat to prior beliefs about how life must have arisen.

In the next chapter, I will address Marshall's claim that building Cambrian animals would not have required large amounts of new genetic information—as I argue in *Darwin's Doubt*—but instead could have been produced by the "rewiring" of preexisting developmental gene regulatory networks.

Notes

1. Charles R. Marshall, "When Prior Belief Trumps Scholarship," *Science*, 341 (September 20, 2013): 1344, http://www.sciencemag.org/content/341/6152/1344.1.s ummary.

2. See Eric H. Davidson, "Evolutionary Bioscience as Regulatory Systems Biology," *Developmental Biology*, 357, no. 1 (September 1, 2011): 40. See also the discussion in *Darwin's Doubt*, 264–70.

3. See, for example, the discussion of the model system *Strongylocentrotus purpuratus* in P. Oliveri, Q. Tu, and E. H. Davidson, "Global Regulatory Logic for Specification of an Embryonic Cell Lineage," *Proceedings of the National Academy of Sciences, USA*, 105, no. 16 (2008): 5955–62. For a diagrammed schematic of the dGRN network of *Strongylocentrotus purpuratus*, see Figure 1D in that article.

4. N. H. Shubin and C. R. Marshall, "Fossils, Genes, and the Origin of Novelty," *Paleobiology*, 26, no. 4, Supplement (December 2000): 335. Emphasis added.

5. Davidson, "Evolutionary Bioscience," 40.

6. See Charles Lyell, *Principles of Geology: Being an Attempt to Explain the Former Changes of the Earth's Surface, by Reference to Causes Now in Operation*. 3 vols. (London: Murray, 1830–33).

11.

NO NEW GENETIC

INFORMATION NEEDED?

Stephen C. Meyer

IN THE PREVIOUS CHAPTER, I ARGUED THAT CHARLES MARSHALL'S review of *Darwin's Doubt* in the journal *Science* illustrates what has become all too common in the defense of contemporary evolutionary theory: the tendency to affirm as true what evolutionary theory requires, even if that contradicts what we know from experiment and observation about how biological systems actually work. Now I will show that in order to rebut the central argument of *Darwin's Doubt*, Marshall must also deny (or at least push from view) what we know about what new forms of animal life *require* as a condition of their existence.

In *Darwin's Doubt*, I argue that intelligent design provides the best explanation for the origin of the genetic (and epigenetic) information necessary to produce the novel forms of animal life that arose in the Cambrian period. To his credit, and unlike other critics of the book, Marshall addresses this, the main argument of the book, and attempts to refute it. To do so, however, he does not show that any of the main materialistic evolutionary mechanisms can produce the information necessary to build the Cambrian animals. Instead, Marshall disputes my claim that significant amounts of new genetic information (and many new protein folds) would have been necessary to build these animals. Specifically, Marshall claims that "rewiring" of dGRNs would have sufficed to produce new animals from a set of preexisting genes. As he argues:

> [Meyer's] case against the current scientific explanations of the relatively rapid appearance of the animal phyla rests on the claim that the origin of new animal body plans requires vast amounts of novel genetic information coupled with the unsubstantiated assertion that this new genetic information must include many new protein folds. In fact, our present understanding of morphogenesis indicates that new phyla were not made by new genes but largely emerged through the rewiring of the gene regulatory networks (GRNs) of already existing genes.[1]

Yet Marshall's understanding of how animal life originated is problematic for several reasons.

First, "rewiring" genetic circuitry would require reconfiguring the temporal and spatial expression of genetic information. Such reconfiguring would entail fixing certain material states and excluding others. Thus, it would constitute an infusion of new information (in the most general theoretical sense) into the biosphere.[2] To see why, consider changing a wiring diagram representing a developmental gene regulatory network. Just altering the diagram representing the network to produce a new diagram representing a new network would require changing the arrangement of "nodes" (representing genes) and "edges" (representing interactions between genes and gene products). Changing the arrangement of these elements in order to produce a new network would constitute adding information to the diagram depicting the system. In the same way, changing the arrangements of genetic elements themselves in an actual network would require informative changes to the arrangement of the network. Thus, Marshall's "rewiring" proposal does not eliminate the need for new information to build the Cambrian animals. Rather, it tacitly invokes additional information of a different, though perhaps partially non-genetic, kind.

In any case, altering the temporal and spatial expression of pre-existing genetic elements assuredly would require the addition of new taxon-specific genes or gene products, and thus, new *genetic* information. Indeed, experiments on dGRNs in modern representatives of the animal phyla show that different organisms use taxon-specific DNA-binding proteins to regulate the expression of genetic data files. For instance, contrary to theoretical expectations,[3] the morphogen Bicoid, essential for normal anterior-posterior body plan specification in *Drosophila*, is found only in the cyclorrhaphan flies.[4] Similarly, the body plan of the freshwater polyp *Hydra* is specified by taxonomically restricted proteins.[5] Nor are these isolated cases. The remarkable disparity of animal morphologies at the macroscopic (i.e., anatomical or body plan) level tends to correspond to differences at *other* levels (i.e., the microscopic or molecular). Moreover, studies of "evo-devo" model systems have repeatedly revealed that the cell and tissue specification programs that generate distinctive animal morphologies depend upon taxon-specific regulatory factors (proteins and RNAs). As Oliveri and Davidson note:

> … the specification apparatus very frequently also includes transcriptional repressors, which, within the specified spatial domain, target key regulatory genes whose expression is required for alternative regulatory states that could have been available to these cells. This is a so-called "exclusion effect," and numerous examples can be found across species…. In each developmental case, *the identity of the specific transcription factor that executes the repression is distinct, as are the specifically excluded target transcription factors.* The design is the same, *the biochemical actors diverse.*[6]

Of course, building these species-specific transcription factors necessary to animal development requires genetic information. And the origin of these proteins in the first place would have required the origin of *new* genetic information.

Second, recent genomic studies of many animals representing phyla that first arose in the Cambrian show that these animals depend upon many *unique* genes not present in any other taxa. Moreover, these genes perform many functions besides just specifying body plan development. These sequences, known as taxonomically restricted or "ORFan" genes, are ubiquitous in all animal life and represent 10 percent or more of the genomes of each species that scientists have investigated.[7]

The presence of ORFan genes in all sequenced present-day animal genomes—and, indeed, in all life[8]—suggests that the genomes of Cambrian animals would have likely contained many ORFan genes as well. That, in turn, suggests that a considerable amount of new genetic information not present in simpler Precambrian organisms would have originated before or during the Cambrian radiation in order to build the unique features of the first Cambrian animals. Moreover, even if some universal Precambrian genome originally contained all the genes that later became taxonomically restricted, whatever process distributed these genes to some lineages, but not others, necessarily involved the addition of new information into the biosphere.[9]

Interestingly, in his review and especially when writing elsewhere, Marshall acknowledges the need for new genes and genetic information in order to produce the Cambrian animals. For example, in a 2006 paper entitled "Explaining the Cambrian 'Explosion' of Animals," he noted: "Animals cannot evolve if the genes for making them are not yet in place. So clearly, developmental/genetic innovation must have played a central role in the radiation."[10] Later in the same paper he argues: "It is also clear that the genetic machinery for making animals must have been in place, at least in a rudimentary way, before they could have evolved."[11] Marshall insists that *Hox* genes, in particular, must have played a necessary causal role in producing the explosion, a point that he also makes in another paper where he explains that developmental considerations "point to the origin of the bilaterian developmental system, including the origin of *Hox* genes, etc., as the primary cause of the 'explosion.'"[12] While in these papers Marshall also emphasizes the importance of rewiring gene regulatory networks to generate new body plans, he clearly acknowledges that new genes would be necessary to produce new animals.

Of course, building the Metazoa (multi-cellular animals) would not have required just new *Hox* genes, ORFan genes, or genes for building new regulatory (DNA-binding) proteins. Instead, the evolutionary process would need to produce a whole range of different proteins necessary to building and servicing the specific forms of animal life that arose in the Cambrian period. In *Darwin's Doubt* I note, for example, that the first arthropods would have likely required genes for building the complex protein lysyl oxidase.[13] Why? Because what we know from studies of modern arthropods shows that this protein is necessary to support the stout body structure of arthropod exoskeletons.[14] Similarly, building Metazoa requires specialized proteins (and metabolic pathways) to produce the kind of extra-cellular matrices that allow developing animals to knit cells into tissues, tissues into organs, and organs and tissues into fully developed animals. Furthermore, different forms of complex animal life exhibit unique cell types and typically each cell type depends upon other specialized or dedicated proteins. As I wrote in *Darwin's Doubt*:

> [New] complex animals [such as arose in the Cambrian period] require more cell types to perform their more diverse functions. Arthropods and mollusks, for example, have dozens of specific tissues and organs, each of which requires "functionally dedicated," or specialized, cell types. These new cell types, in turn, require many new and specialized proteins. An epithelial cell lining a gut or intestine, for example, secretes a specific digestive enzyme. This enzyme requires structural proteins to modify its shape and regulatory enzymes to control the secretion of the digestive enzyme itself. Thus, building novel cell types typically requires building novel proteins, which requires assembly instructions for building proteins—that is, genetic information.[15]

Thus, our present observations of animals representing the phyla that first arose in the Cambrian show that these animals would have needed many specialized proteins: proteins for building extracellular matrices or exoskeletons, for facilitating adhesion, for regulating development, for building specialized tissues or structural parts of specialized organs, for servicing gut cells, for producing eggs and sperm, and many other distinctive functions and structures of individual metazoans. Obviously, these proteins would have had to arise sometime in the history of life. Since most of the Metazoa first arose in the Cambrian explosion, it is reasonable to infer that the proteins necessary to sustain those forms of animal life also arose around that time or just before.

Although Marshall characterizes my claim that new Cambrian animals would have required new genetic information and new protein folds as "unsubstantiated," he doesn't actually dispute the need for genetic information in order to build the proteins required by each new form of metazoan life. Instead, he only seems to dispute that all that information arose *during* the Cambrian explosion itself. Indeed, in both his technical publications and his review of *Darwin's Doubt*, Marshall simply assumes that most of the genetic information necessary to build the Cambrian animals already existed *before* the Cambrian explosion. In fact, he seems to presuppose the existence of what Susumu Ohno called a "pananimalian genome,"[16] a nearly complete set of the genes necessary to build Cambrian animals within some phenotypically simpler, *ur-*metazoan ancestor. Thus, he states that the new animal phyla "emerged through the rewiring of the gene regulatory networks (GRNs) *of already existing genes.*"[17] The article "The Causes of the Cambrian Explosion," which accompanies Marshall's review of my book in *Science,* also presupposes such a universal gene toolkit and suggests that it might have arisen 100 million years or more before the explosion of animal life in the Cambrian period.[18]

Nevertheless, this question-begging assumption does not solve the central problem posed by *Darwin's Doubt*—that of *the origin* of the genetic (and epigenetic) information necessary to produce the Cambrian animals. It merely pushes the problem back several tens or hundreds of millions of years, assuming that such a universal genetic toolkit ever existed. (Marshall also makes no attempt to rebut my argument about the inability of the mutation/selection mechanism to generate new epigenetic information, a problem that has led other prominent evolutionary biologists to express skepticism about the adequacy of the neo-Darwinian mechanism.[19]) In any case, Marshall does not explain how the neo-Darwinian mechanism could have overcome the combinatorial search problem described in *Darwin's Doubt* to produce even the new *genetic* information necessary to build new proteins and Cambrian animals.

Readers of the book will recall my discussion, in Chapters 9 and 10, of recent mutagenesis experiments. These experiments have established the *extreme* rarity of functional genes and proteins among the many (combinatorially) possible ways of arranging nucleotide bases or amino acids within their corresponding "sequence spaces." Readers will also recall that the rarity of functional genes and proteins within sequence space makes it overwhelmingly more likely than not that a series of random mutation searches will *fail* to generate even a single new gene or protein fold within available evolutionary time. This extreme rarity also helps to explain why mathematical biologists, using standard population genetics models, are calculating exceedingly long waiting times (well in excess of available evolutionary time) for the production of new genes and proteins when producing such genes or proteins requires even a few coordinated mutations.[20]

For these reasons, defining the Cambrian explosion as a 25-million-year event, as Marshall does, instead of a 10-million-year event, as many other Cambrian experts do (and as I do in *Darwin's Doubt*), makes no appreciable difference in solving the problem of the origin of genetic information—such is the extreme rarity of functional bio-macromolecules within their relevant sequence spaces. Nor, for that matter, does positing the origin of a *complete set* of genes (that is, many more than just one) for building all the Cambrian animals 100 million years before the Cambrian explosion. That merely pushes the problem back and raises other problems such as (a) explaining exactly what selective advantage all these genes for building new animals would have had before they were actually used to build the diverse animals that arose in the Cambrian and (b) how the maintenance of this overly complex genome could have avoided exacting a huge energetic and fitness cost on its host organism, and thus the effects of purifying selection over 100 million years of evolutionary time.

In any case, the experimentally based calculations in *Darwin's Doubt* show that neither ten million nor several hundred million years would afford enough opportunities to produce the genetic information necessary to build even a single novel gene or protein, let alone all the new genes and proteins needed to produce new animal forms. Indeed, neither stretch of time is sufficient to allow the mutation/selection process to search more than a tiny fraction of the relevant sequence spaces. Marshall's review does not even allude to a solution to this long-standing mathematical,[21] and now experimentally based,[22] challenge to the efficacy of the neo-Darwinian mechanism. Instead, his proposal merely presupposes the prior existence of the genetic information necessary to produce the Cambrian animals.

In the next chapter, I'll address Marshall's claim that the positive argument for intelligent design that I make in the book actually constitutes a wholly negative or critical "God of the gaps" argument.

Notes

1. Charles R. Marshall, "When Prior Belief Trumps Scholarship," *Science* 341, no. 6152 (September 20, 2013): 1344.

2. See Claude Shannon, "A Mathematical Theory of Communication," *Bell System Technical Journal*, 27 (1948): 370–423, 623–29.

3. Nicolas Rasmussen, "A New Model of Developmental Constraints as Applied to the *Drosophila* System," *Journal of Theoretical Biology* 127, no. 3 (August 1987): 271–99.

4. David Rudel and Ralf Sommer, "The evolution of developmental mechanisms," *Developmental Biology* 264, no. 1 (December 2003): 15–37; From p. 25: "Phylogenetic evidence suggests that *bcd* may be a new innovation in the anterior positional information gene network during the evolution of Dipterans. Despite repeated attempts, it has not been possible to clone *bcd* homologues outside of the Cyclorraphan flies (Stauber et al., 1999). Additionally, *bcd* is not present in the *Antennapedia* complex of the flour beetle *Tribolium castaneum* (Brown et al., 2002). This has caused speculation that *bcd* may have evolved late in the evolution of the Dipterans."

5. Konstantin Khalturin et al., "A Novel Gene Family Controls Species-Specific Morphological Traits in Hydra," *PLOS Biology* 6, no. 11 (November 2008): pe278.

6. Paolo Oliveri and Eric H. Davidson, "Built to Run, Not Fail," *Science* 315, no. 5818 (March 16, 2007), 1510–11. Emphasis added.

7. Amanda K. Gibson et al., "Why so many unknown genes? Partitioning orphans from a representative transcriptome of the lone star tick *Amblyomma americanum*," *BMC Genomics* 14 (February 2013): 135.

8. Konstantin Khalturin et al., "More than just orphans: are taxonomically-restricted genes important in evolution?" *Trends in Genetics* 25, no. 9 (September 2009): 404–13.

9. Marshall seems to have an idiosyncratic view of animal evolution, depicting the evolution of animals as a reductive process in which pre-existing genetic information from a universal Precambrian gene set is selectively lost to some lineages but not to others. This contrasts markedly with a more standard neo-Darwinian view in which form and information gradually accumulate over time.

10. Charles R. Marshall, "Explaining the Cambrian 'Explosion' of Animals," *Annual Review of Earth and Planetary Sciences* 34 (2006): 355–84.

11. Ibid.

12. Charles R. Marshall, "Nomothetism and Understanding the Cambrian 'Explosion,'" *Palaios* 18, no. 3 (June 2003): 195–96.

13. See Stephen C. Meyer, *Darwin's Doubt: The Explosive Origin of Animal Life and the Case for Intelligent Design* (New York: HarperOne, 2013), 191.

14. Susumu Ohno, "The notion of the Cambrian pananimalia genome," *Proceedings of the National Academy of Sciences, USA* 93, no. 16 (August 6, 1996): 8475–78.

15. *Darwin's Doubt*, 162.

16. Susumu Ohno, "The notion of the Cambrian pananimalia genome."

17. Marshall, "Prior Belief," 1344. Emphasis added.

18. M. Paul Smith and David A. T. Harper, "Causes of the Cambrian Explosion," *Science* 341 no. 6152 (September 20, 2013): 1355–56. (Smith and Harper propose "an apparent >100-million-year gap between the evolutionary innovation and its consequences.")

19. Gerd B. Müller and Stuart A. Newman, "Origination of Organismal Form: The Forgotten Cause in Evolutionary Theory," in *Origination of Organismal Form: Beyond the Gene in Developmental and Evolutionary Biology*, edited by Gerd B. Müller and Stuart A. Newman (Cambridge, MA: MIT Press, 2003), 7–8.

20. See *Darwin's Doubt*, Chapters 9–12.

21. Murray Eden, "Inadequacies of Neo-Darwinian Evolution as a Scientific Theory" in *Mathematical Challenges to the Neo-Darwinian Interpretation of Evolution*, edited by P. S. Moorhead and M. M. Kaplan, 9–11. Wistar Institute Symposium Monograph (New York: Liss, 1967); Marcel Schützenberger, "Algorithms and the Neo-Darwinian Theory of Evolution." Also in *Mathematical Challenges to Neo-Darwinian Theory of Evolution*, 73–80.

22. John Reidhaar-Olson and Robert Sauer, "Functionally Acceptable Solutions in Two Alpha-Helical Regions of Lambda Repressor," *Proteins: Structure, Function, and Genetics* 7, no. 4 (1990): 306–16; Douglas D. Axe, "Estimating the Prevalence of Protein Sequences Adopting Functional Enzyme Folds," *Journal of Molecular Biology* 341, no. 5 (August 27, 2004), 1295–1315.

12.

THE GOD-OF-THE-GAPS FALLACY?

Stephen C. Meyer

IN MY PREVIOUS DISCUSSION (CHAPTERS 10 AND 11) OF CHARLES Marshall's review in *Science* of my book *Darwin's Doubt*, I responded to Marshall's claim that building Cambrian animals would *not* have required large amounts of new genetic information, but instead could have been produced by "rewiring" of preexisting developmental gene regulatory networks.[1] I showed that Marshall's proposal for "rewiring" gene regulatory networks would itself require an infusion of new information into the biosphere and that Marshall had, in any case, merely pushed the problem of the origin of genetic information back several tens or hundreds of millions of years by presupposing a pre-existing set of genes for building the Cambrian animals in some hypothetical Precambrian ancestor.

I turn now to his claim that the book's argument for intelligent design represents a purely negative "god of the gaps" argument. Marshall writes in his review:

> Meyer's scientific approach is purely negative. He argues that paleontologists are unable to explain the Cambrian explosion, thus opening the door to the possibility of a designer's intervention. This, despite his protest to the contrary, is a (sophisticated) "god of the gaps" approach, an approach that is problematic in part

because future developments often provide solutions to once apparently difficult problems.[2]

I appreciate Marshall's compliment about the sophistication with which I allegedly marshal this fallacious form of argumentation. Nevertheless, his characterization of my argument is entirely inaccurate. First, although I do acknowledge in the last chapter of *Darwin's Doubt* that the case for intelligent design has implications that are friendly to theistic belief (since all theistic religions affirm that the universe and life are the product of a designing intelligence), the scientific argument that I make does not attempt to establish the existence of God. I attempt merely to show that key features of the Cambrian animals (and the pattern of their appearance in the fossil record) are best explained by a designing intelligence—a conscious rational agency or a mind—of some kind. Thus, my argument does not qualify as a God-of-the-gaps argument for the simple reason that the argument does not attempt to establish the existence of God.[3]

But let's set aside what Marshall might regard as a trivial distinction about what I claim—or rather don't claim—to have established about the identity of the designing intelligence responsible for life. By claiming that my approach is a purely negative one based solely upon "gaps" in our knowledge or in the evolutionary account of the Cambrian explosion, Marshall implies that *Darwin's Doubt* makes a fallacious kind of argument known to logicians as an "argument from ignorance." Arguments from ignorance occur when evidence against a proposition X is offered as the sole (and conclusive) grounds for accepting some alternative proposition Y. Arguments from ignorance make an obvious logical error. They omit a necessary kind of premise, a premise providing positive support for the conclusion, not just negative evidence against an alternative conclusion. In an explanatory context, arguments from ignorance have the form:

Premise One: Cause X cannot produce or explain evidence E.

Conclusion: Therefore, cause Y produced or explains E.

Critics of intelligent design often claim that the case for intelligent design commits this fallacy.[4] They claim that design advocates use our present ignorance of any materialistic cause of specified or functional information (for example) as the sole basis for inferring an intelligent cause for the origin of such information in biological systems. For example, Michael Shermer represents the case for intelligent design as follows: "intelligent design… argues that life is too specifically complex (complex structures with specific functions like DNA)… to have evolved by natural forces. Therefore, life must have been created by… an intelligent designer."[5] In short, Shermer claims that ID proponents argue as follows:

Premise: Materialistic causes or evolutionary mechanisms cannot produce novel biological information.

Conclusion: Therefore, an intelligent cause produced specified biological information.

Marshall echoes Shermer's criticism. But the inference to design as developed in *Darwin's Doubt* does not commit this fallacy. True, the book does offer several evidentially based (and mathematically rigorous) arguments *against* the creative power of the mutation/natural selection mechanism (none of which Marshall refutes). And clearly, this lack of knowledge of any adequate materialistic evolutionary cause of, for example, the biological information necessary to produce novel forms of animal life, does provide *part* of the grounds for the inference to intelligent design presented in *Darwin's Doubt*. (However, it is probably more accurate to characterize this "absence of knowledge" as knowledge of inadequacy, since it derives from a thorough assessment of causal powers—and limitations—of various materialistic evolutionary mechanisms.) In any case, the argument presented in the book is not, as Marshall claims, a "purely negative" and, therefore, fallacious argument based on the inadequacy of various materialistic evolutionary mechanisms (or gaps in our knowledge).

Instead, the book makes a *positive* case for intelligent design as an inference to the best explanation for the origin of the genetic (and epigenetic) information necessary to produce the first forms of animal life (as well as other features of the Cambrian animals such as the presence of genetic regulatory networks that function as integrated circuits during animal development). It advances intelligent design as the best explanation not only because many lines of evidence now cast doubt on the creative power of unguided evolutionary mechanisms, but also because of our *positive*, experience-based knowledge of the powers that intelligent agents have to produce digital and other forms of information as well as integrated circuitry. As I argue in Chapter 18 of *Darwin's Doubt*:

> Intelligent agents, due to their rationality and consciousness, have demonstrated the power to produce specified or functional information in the form of linear sequence-specific arrangements of characters. Digital and alphabetic forms of information routinely arise from intelligent agents. A computer user who traces the information on a screen back to its source invariably comes to a mind—a software engineer or programmer. The information in a book or inscription ultimately derives from a writer or scribe. Our experience-based knowledge of information flow confirms that systems with large amounts of specified or functional information invariably originate from an intelligent source. The generation of functional information is "habitually associated with conscious activity." Our uniform experience confirms this obvious truth.[6]

Thus, the inadequacy of proposed materialistic evolutionary causes or mechanisms forms only part of the basis of the argument *for* intelligent design. We also *know* from broad and repeated experience that intelligent agents can and do produce information-rich systems and integrated circuitry. We have positive experience-based knowledge of a cause sufficient to generate new specified information and integrated circuitry, namely, intelligence. We are not ignorant of how information or circuitry arises. We know that conscious, rational agents can create such information-rich structures and systems. Indeed, whenever large amounts of specified or functional information are present in an artifact or entity whose causal story is known, invariably creative intelligence—intelligent design—played a role in the origin of that entity. Thus, when we encounter a large discontinuous increase in the functional information content of the biosphere as we do in the Cambrian explosion, we may infer that a purposive intelligence operated in the history of life to produce the functional information necessary to generate those forms of animal life.

Instead of exemplifying a fallacious form of argument in which design is inferred solely from a negative premise, the argument for intelligent design formulated in *Darwin's Doubt* takes the following form:

> **Premise One:** Despite a thorough search and evaluation, no materialistic causes or evolutionary mechanisms have demonstrated the power to produce large amounts of specified or functional information (or integrated circuitry).

> **Premise Two:** Intelligent causes have demonstrated the power to produce large amounts of specified/functional information (and integrated circuitry).

> **Conclusion:** Intelligent design constitutes the best, most causally adequate explanation for the specified/functional information (and circuitry) that was necessary to produce the Cambrian animals.

The second affirmative premise in this argument makes clear that the design argument in *Darwin's Doubt* does not constitute an argument from ignorance, nor is it a "purely negative" argument. Indeed, in addition to showing that various materialistic causes lack demonstrated causal adequacy, my argument for intelligent design also affirms the demonstrated causal adequacy of an alternative cause, namely, intelligence. My argument does not omit a premise providing positive evidence or reasons for preferring an alternative non-materialistic cause or proposition.

In fact, the argument for intelligent design developed in *Darwin's Doubt* constitutes an "inference to the best explanation" based upon our best available knowledge.[7] As I note in Chapter 17 of the book, to establish an explanation as best, a historical scientist must cite positive evidence for the causal adequacy of a proposed cause. Unlike an argument from ignorance, an inference to the best explanation does not assert the adequacy of one causal explanation merely on the basis of the inadequacy of some other causal explanation. Instead, it asserts the superior explanatory power of a proposed cause based upon its established—its *known*—causal adequacy, *and* based upon a lack of demonstrated efficacy, despite a thorough search, of any other adequate cause. The inference to design, therefore, depends on present *knowledge* of the causal powers of various materialistic entities and processes (inadequate) and intelligent agents (adequate).

Formulated this way, the argument to design from biological information also exemplifies the standard uniformitarian canons of method employed within the historical sciences. The uniformitarian method affirms that "the present is the key to the past."[8] In particular, the principle specifies that our *knowledge* of present cause-effect relationships should govern how we assess the plausibility of inferences we make about the causes of events in the remote past. Determining which explanation, among a set of competing alternatives, constitutes the best depends on *knowledge* of the causal powers of the possible explanatory entities, knowledge that we acquire through our repeated observation and experience of the cause-and-effect patterns of the world.[9] Such knowledge, not ignorance, undergirds my inference to intelligent design from the features of the Cambrian animals. It no more constitutes an argument from ignorance than any other well-grounded inference in geology, archeology or paleontology—where present knowledge of cause-effect relationships guides the inferences that scientists make about events in the past.

Marshall treats my argument as a "god-of-the-gaps" argument not because it actually has the form of a logically fallacious "argument from ignorance," but because he tacitly presupposes that materialistic causes will ultimately suffice to explain all events in the history of life and that *only* such explanations count as *scientific* explanations. Yet we know from our uniform and repeated experience that some types of phenomena—in particular, information-rich sequences and systems—do not arise from mindless, materialistic processes. For just this reason, no rational person would, for example, insist that the inscriptions on the Rosetta

Stone in the British museum *must have* been produced by purely materialistic causes such as wind and erosion.

Yet Marshall and many other evolutionary biologists maintain an *a priori* commitment to purely materialistic explanation for all events in the history of life, even events such as the Cambrian explosion that necessarily involve the generation of massive amounts of new functional information. By privileging prior commitments to a purely materialistic account of evolutionary history over our present knowledge of cause and effect—in particular, our knowledge that intelligent agents, and only intelligent agents, produce information-rich systems and structures—Marshall and others disregard the methodological imperatives of the uniformitarian method, privileging what we don't observe (about what happened in the evolutionary past) over what we do observe (the causal powers of various entities and processes). Thus, ironically, Marshall does precisely what he thinks he sees me doing: he allows his own prior commitment to a belief system—evolutionary materialism—to trump objective analysis of the observational evidence.

In Chapter 14, I will conclude my response to Marshall with a postscript on two other substantive, but minor, criticisms of *Darwin's Doubt*.

Notes

1. See Stephen C. Meyer, "To Build New Animals, No New Genetic Information Needed? More in Reply in Charles Marshall," *Evolution News & Views*, October 7, 2013, http://www.evolutionnews.org/2013/10/to_build_new_an077541.html; Stephen C. Meyer, "When Theory Trumps Observation: Responding to Charles Marshall's Review of *Darwin's Doubt*," *Evolution News & Views*, October 2, 2013, http://www.evolutionnews.org/2013/10/when_theory_tru077391.html.

2. Marshall claims *Darwin's Doubt* presents "a (sophisticated) 'god of the gaps' approach." Charles R. Marshall, "When Prior Belief Trumps Scholarship," *Science* 341, no. 6152 (September 20, 2013): 1344.

3. I explain this point in more detail in my previous book, *Signature in the Cell*: "The theory of intelligent design does not claim to detect a supernatural intelligence possessing unlimited powers. Though the designing agent responsible for life may well have been an omnipotent deity, the theory of intelligent design does not claim to be able to determine that. Because the inference to design depends upon our uniform experience of cause and effect in this world, the theory cannot determine whether or not the designing intelligence putatively responsible for life has powers beyond those on display in our experience. Nor can the theory of intelligent design determine whether the intelligent agent responsible for the information in life acted from the natural or the

"supernatural" realm. Instead, the theory of intelligent design merely claims to detect the action of some intelligent cause (with power at least equivalent to that of intelligent causes known from experience) and affirms this because we know from experience that only conscious, intelligent agents produce large amounts of specified information. The theory of intelligent design does not claim to be able to determine the identity or any other attributes of that intelligence, even if philosophical deliberation or additional evidence from other disciplines may provide reasons to consider, for example, a specifically theistic design hypothesis." (*Signature in the Cell*, 428–429)

4. John S. Wilkins and Wesley R. Elsberry, "The Advantages of Theft over Toil: The Design Inference and Arguing from Ignorance," *Biology & Philosophy* 16, no. 5 (November 2001): 711–24.

5. Michael Shermer, "ID Works in Mysterious Ways," *The Ottawa Citizen*, July 9, 2008, http://www.canada.com/story.html?id=711a0b47-29d5-426d-a273-a270817b000e.

6. Stephen C. Meyer, *Darwin's Doubt: The Explosive Origin of Animal Life and the Case for Intelligent Design* (New York: HarperOne, 2013), 360, citing Henry Quastler, *The Emergence of Biological Organization* (New Haven, CT: Yale Univ. Press, 1964), 16. Emphasis added.

7. Peter Lipton, *Inference to the Best Explanation* (London: Routledge, 1991), 32–88.

8. This principle is based upon the arguments of Charles Lyell, *Principles of Geology: Being an Attempt to Explain the Former Changes of the Earth's Surface, by Reference to Causes Now in Operation*. 3 vols. (London: Murray, 1830–33).

9. Lipton, *Inference to the Best Explanation*, 32–88; Stephen C. Meyer, "The Scientific Status of Intelligent Design: The Methodological Equivalence of Naturalistic and Non-Naturalistic Origins Theories," in *Science and Evidence for Design in the Universe* (San Francisco: Ignatius Press, 2000), 151–212; Stephen C. Meyer, "The Demarcation of Science and Religion," in Gary B. Ferngren, ed., *The History of Science and Religion in the Western Tradition: An Encyclopedia* (New York: Garland Pub., 2000), 17–23; Elliott Sober, *The Philosophy of Biology* (Boulder, CO: Westview Press, 1993); Stephen C. Meyer, *Of Clues and Causes: A Methodological Interpretation of Origin of Life Studies*, PhD dissertation, University of Cambridge, 77–140.

13.

SMALL SHELLY FOSSILS AND

THE CAMBRIAN EXPLOSION

Casey Luskin

IN HIS REVIEW OF *DARWIN'S DOUBT* IN THE JOURNAL *SCIENCE*, PALE-
ontologist Charles Marshall claims that Stephen Meyer "completely
omits mention of the Early Cambrian small shelly fossils," which he
thinks causes Meyer "to exaggerate the apparent suddenness of the
Cambrian explosion."[1] Yet on both points Marshall is wrong. Meyer
does not fail to mention the small shelly fossils and he does not exaggerate the brevity of the Cambrian explosion.

Meyer discusses the small shelly fossils in *Darwin's Doubt*. He writes
as follows:

> The Cambrian period 543 mya is marked by the appearance of
> small shelly fossils consisting of tubes, cones, and possibly spines
> and scales of larger animals. These fossils, together with trace
> fossils, gradually become more abundant and diverse as one
> moves upward in the earliest Cambrian strata (the Manykaian
> Stage, 543–530 mya).[2]

Nevertheless, although Meyer discusses the small shelly fossils, he
does not treat them as a solution to the problem of the explosion of morphological novelty that arises later in the Cambrian period. The small
shelly fossils appear in the fossil record at the base of the Cambrian

period about 542–543 million years ago. The main pulse of morphological innovation that Cambrian paleontologists commonly refer to as
the "Cambrian explosion" first begins about 530 million years ago and
then lasts about 10 million years through the Tommotian and Atdabanian stages of the Cambrian period. During the first 5–6 million year
stage (the Tommotian) of the explosion, between 14–16 novel phyla first
appear in the fossil record.[3] Without actually asserting that the small
shelly fossils somehow explain the subsequent explosion of all these
novel forms of animal life (or even that the small shelly fossils represent
ancestors to all, or some, of these forms), Marshall faults Meyer for not
treating them as part of the Cambrian explosion.

But does Meyer's decision not to treat them as clear ancestors of
the later forms mean that that he exaggerated the brevity of the explosion and, in so doing, overlooked a possible explanation for the missing
ancestral fossils to the animals that arise in the crucial Tommotian and
Atdabanian periods?

It doesn't—as Marshall's own technical writing has made clear. For
example, in a 2006 paper in *Annual Reviews of Earth and Planetary Sciences*, Marshall acknowledges that these fossils are of unclear evolutionary affinities and importance. He calls them "largely problematic fossils"
that are "hard to diagnose, even at the phylum level."[4] Figure 1 of his
paper portrays them as apparently disconnected from the later radiation
of Cambrian animals. This impression is reinforced in the text of his
article where he notes that the small shelly fossils for the most part are
"problematic" organisms of unknown classification:

> While many represent individual animals, others represent in-
> dividual components of the armor of much larger animals (Con-
> way Morris & Peel 1995). Some of the described genera belong
> to known phyla such as Brachiopoda and Mollusca. However,
> many are problematic, including the cambroclaves, coelosclerito-
> pherans, cribricyatheans, machaeridians, tommotiids, as well as
> a diverse array of incertae sedis [fossils of uncertain taxonomic
> placement].[5]

Other authorities agree that these small shelly fossils [SSFs] are of unclear evolutionary significance and affinity. In his book *On the Origin of Phyla*, James Valentine argues that the SSFs "are very difficult indeed to interpret."[6] Valentine's 2013 book, *The Cambrian Explosion*, co-written with Douglas Erwin, notes that "many SSFs are still poorly understood."[7] Simon Conway Morris found them so unimportant that he does not mention them in either of his authoritative books on the Cambrian explosion (*Crucible of Creation* or *Life's Solution*).

Nevertheless, Marshall faults Meyer for failing to mention the SSFs and claims this alleged oversight resulted in his understating the length of the Cambrian explosion:

> Meyer completely omits mention of the Early Cambrian small shelly fossils and misunderstands the nuances of molecular phylogenetics, both of which cause him to exaggerate the apparent suddenness of the Cambrian explosion.

Now Marshall never mentions any specific errors in Meyer's treatment of molecular phylogenetics, so we must await his further critique on that subject. But what about the claim that *Darwin's Doubt* exaggerated the brevity of the Cambrian explosion? Should Meyer have included the appearance of the early Cambrian small shelly fossils as part of the explosion when he estimated the length of that event? Not according to a very recent paper by Marshall himself. In 2010, Marshall co-wrote with James Valentine in the journal *Evolution*:

> By the beginning of the Cambrian Period, near 543 million years ago, a few kinds of "small shelly" fossils are found, <2mm in largest dimension. The small shellys rose to a peak in abundance and diversity during the period from 530 to 520 million years ago, when representatives of living phyla are found among them. During that same period, a *chiefly larger-bodied invertebrate fauna* of up to a dozen phyla, and including many soft-bodied forms, is also first represented by fossils. This geologically abrupt appearance of fossils representing *quite disparate bodyplans* of many living metazoan phyla is termed the Cambrian explosion.[8]

Let's consider the construction of this paragraph, in which Marshall explains the length of the Cambrian explosion in relation to the small shelly fossils. Starting at the end, Marshall and Valentine equate "the Cambrian explosion" with the "geologically abrupt appearance of fossils representing quite disparate body plans." They further identify this period with "that same period" wherein "a chiefly larger-bodied invertebrate fauna of up to a dozen phyla, and including many soft-bodied forms, is also first represented by fossils." Marshall and Valentine also equate that period of time with "the period from 530 to 520 million years ago" and distinguish it from the earlier time in which the first small shelly fossils arose. Thus, according to Marshall—in a co-authored technical paper written in 2010—the Cambrian explosion does not begin with the first appearance of the small shelly fossils 543 million years ago, or during the earliest part of the Cambrian period. Rather, he and fellow paleontologist James Valentine affirm that the explosion begins about 530 million years ago and lasted to about 520 million years ago—a date consistent with what Valentine has written elsewhere, including in his recent book with Erwin that Marshall cites approvingly in his review of Meyer. There he writes:

> [A] great variety and abundance of animal fossils appear in deposits dating from a geologically brief interval between about 530 to 520 Ma, early in the Cambrian period. During this time, nearly all the major living animal groups (phyla) that have skeletons first appeared as fossils (at least one appeared earlier). Surprisingly, a number of those localities have yielded fossils that preserve details of complex organs at the tissue level, such as eyes, guts, and appendages. In addition, several groups that were entirely soft-bodied and thus could be preserved only under unusual circumstances also first appear in those faunas. Because many of those fossils represent complex groups such as vertebrates (the subgroup of the phylum Chordata to which humans belong) and arthropods, it seems likely that all or nearly all the major phylum-level groups of living animals, including many small soft-bodied groups that we do not actually find as fossils, had appeared by the end of the early Cambrian. This geologically abrupt and spectacular record of early animal life is called the Cambrian explosion.[9]

Thus, by Marshall's own admission, (a) the appearance of small shelly fossils around 543 million years ago does *not* mark the beginning of the Cambrian explosion, and (b) the Cambrian explosion should be dated to 530 to 520 million years ago, when we see the "abrupt appearance" of many disparate body plans, long after the small shellies appear. This means that Marshall has acknowledged in print that the "Cambrian explosion" itself lasted only about 10 million years—just as Meyer says in *Darwin's Doubt*. Indeed, Marshall and Valentine write that SSFs appear long before the primary explosive radiation of Cambrian animals and they affirm a 10-million-year duration for the Cambrian explosion. In response to Nick Matzke, I documented many scientific papers written by other Cambrian experts that also assign an approximately 10-million-year period for the main pulse of morphological innovation that paleontologists typically call the Cambrian explosion.[10] So here again we see one of Meyer's critics criticizing Meyer for holding a position[11] about a factual matter that leading Cambrian paleontologists also hold—in this case, a position that Marshall himself has sometimes publicly affirmed.

It's revealing that Marshall doesn't actually claim that the small shelly fossils solve the problem of the explosion of morphological novelty that occurs later in the Cambrian period. Instead, he seems content to use the small shelly fossils as a rhetorical cudgel, knowing, I suspect, that these fossils do little if anything to diminish the real problem of morphological novelty that makes the subsequent stages of the Cambrian period so vexing from a Darwinian point of view.

Notes

1. Charles R. Marshall, "When Prior Belief Trumps Scholarship," *Science* 341, no. 6152 (September 20, 2013): 1344.

2. Stephen C. Meyer, *Darwin's Doubt: The Explosive Origin of Animal Life and the Case for Intelligent Design* (New York: HarperOne, 2013–14), 425 (2013 hardback) and 460 (2014 paperback). That is the endnote 39 for chapter 4, "The *Not* Missing Fossils."

3. Erwin et al., "The Cambrian Conundrum: Early Divergence and Later Ecological Success in the Early History of Animals," *Science* 334, no. 6059 (November 25, 2011): 1091–1097; Samuel A. Bowring et al., "Calibrating Rates of Early Cambrian Evolution," *Science* 261, no. 5126 (September 3, 1993): 1293–1298.

4. Charles R. Marshall, "Explaining the Cambrian 'Explosion' of Animals," *Annual Review of Earth and Planetary Sciences* 34 (May 2006), 360.

5. Ibid., 360.

6. James W. Valentine, *On the Origin of Phyla* (Chicago: University of Chicago, 2004), 304.

7. Douglas Erwin and James Valentine, *The Cambrian Explosion: The Construction of Animal Biodiversity* (Greenwood Village, CO: Roberts and Co., 2013), 151.

8. Charles R. Marshall and James W. Valentine, "The Importance of Preadapted Genomes in the Origin of the Animal Bodyplans and the Cambrian Explosion," *Evolution* 64, no. 5 (May 2010): 1189–1201. Emphasis added.

9. Erwin and Valentine, *The Cambrian Explosion*, 5.

10. See Chapter 6 in this book, "How 'Sudden' Was the Cambrian Explosion?" by Casey Luskin.

11. Meyer equates the Cambrian explosion with the most explosive period of the Cambrian radiation (as most Cambrian experts do) in which the vast majority of the higher taxa arose. As he argues, the re-dating of critical Cambrian strata in 1993 established that the strata documenting the first appearance of the majority of the Cambrian phyla and classes took place within a 10-million-year period—a period he calls "the explosion of novel Cambrian animal forms." (*Darwin's Doubt*, 71) As Meyer writes: "these studies [i.e., radiometric analyses of zircon crystals in Siberian rocks] also suggested that the explosion of novel Cambrian animal forms" took about 10 million years. (*Darwin's Doubt*, 71)

14.

MORE ON SMALL SHELLY FOSSILS

AND THE CAMBRIAN EXPLOSION

Stephen C. Meyer

IN MY CHAPTERS REPLYING TO CHARLES MARSHALL'S REVIEW IN *Science* of *Darwin's Doubt*, I've responded to his critiques of the main argument of the book—in particular, to (a) his claim that the Cambrian explosion would not have required a significant increase in new genetic information and (b) his claim that my positive argument for intelligent design, based on the need for an increase in genetic (and other forms of biological) information, represents a purely negative "god of the gaps" argument.

In this concluding response, I will address two other substantive, but minor, criticisms that Marshall makes of *Darwin's Doubt*. In his review in *Science*, Marshall claims that the book fails to discuss the small shelly fossils that arise at the base of the Cambrian period, and thereby exaggerates the brevity of the Cambrian explosion,[1] treating it as a 10-million-year event, rather than a roughly 25-million-year one, as Marshall sometimes (but not always) does.

The first of these two claims is false. *Darwin's Doubt* discusses the small shelly fossils in the following paragraph:

The Cambrian period 543 mya is marked by the appearance of small shelly fossils consisting of tubes, cones, and possibly spines and scales of larger animals. These fossils, together with trace fossils, gradually become more abundant and diverse as one moves upward in the earliest Cambrian strata (the Manykaian Stage, 543–530 mya).[2]

Of course, Marshall in his review implies that *Darwin's Doubt* should have treated the first appearance of the small shelly fossils as part of the Cambrian explosion. The main pulse of morphological innovation that many Cambrian paleontologists designate as the explosion took place between 530 and 520 million years ago. Marshall faults *Darwin's Doubt* for failing to include the first appearance of the small shelly fossils beginning 12–13 million years earlier (543–542 million years ago) as the beginning of the explosion, a decision that would imply a 22–23 million-year event, rather than a 10-million-year event.

Readers should note that Casey Luskin has already extensively rebutted the claim that *Darwin's Doubt* exaggerates the brevity of the Cambrian explosion. (See Chapter 6.) I have done the same in my reply to John Farrell published in *National Review*[3] and as Chapter 21 in this book. As I explain in my response to Farrell, *Darwin's Doubt* "affirms the widely accepted figure among Cambrian paleontologists of about 10 million years for the main pulse of morphological innovation in the Cambrian period that paleontologists typically designate as 'the explosion.'" Luskin also documents that this figure is widely accepted among many Cambrian experts, including Valentine and Erwin, whom Marshall cites affirmatively in his review.

As Luskin shows in Chapter 13, Marshall *himself*, like many other Cambrian experts, does not regard the small shelly fossils as obviously ancestral to most of the animals that arise in the main explosive period of the Cambrian radiation. In one 2006 paper he depicts them as (apparently) disconnected from the later more significant pulses of morphological innovation.[4] In fact, Marshall notes repeatedly that the small shelly fossils are "largely problematic" and "hard to diagnose even at the phylum level."[5] As Luskin points out, in a technical article published in 2010, Marshall specifically excludes the small shelly fossils from the 10-million-year "geologically abrupt appearance of fossils representing quite disparate body plans" that he and co-author James Valentine designate "as the Cambrian explosion."

In any case, treating the first appearance of the small shelly fossils as the beginning of the Cambrian explosion does little to explain the main pulse of the morphological innovation that occurs later during the 10-million-year window. As I acknowledge in *Darwin's Doubt*, it is entirely possible to assign a different duration to the "Cambrian explosion" depending upon how many separate paleontological events scientists choose to include within that designation. Nevertheless, quibbling of that sort reduces the debate to one of semantics. The key question is not how many different events should be included within the designation "Cambrian explosion." Nor is it about the total amount of time that some arbitrarily designated series of separate paleontological events covers. Instead, the key question is what *caused* the discontinuous appearance of morphological novelty within specific, and measurably narrow, windows of geological time—whatever we choose to call them. Thus, *Darwin's Doubt* focuses on the crucial Tommotian and Atdabanian stages of the Cambrian explosion—where 13 to 16 new animal phyla arose within a 5 to 6 million year window—as a defining challenge to the efficacy of the neo-Darwinian mechanism. Marshall doesn't explain how the origin of the small shelly fossils diminishes that problem.

Moreover, as I have discussed in response to Marshall (and in Chapters 10 and 12 of *Darwin's Doubt*), even a duration of 25 million years would not appreciably diminish the problem facing contemporary evolutionary theory. In the first place, 25 million years would not provide enough opportunities for the mutation/selection process to search more than a tiny fraction of the relevant sequence space necessary to produce even a single new gene or functional protein.[6] Second, the calculated waiting times required to evolve multi-mutation features also suggest that even pushing the beginning of the Cambrian explosion back to the first appearance of the small shelly fossils, as Marshall suggests we should, does not provide enough time for many complex biological features to evolve.[7] Marshall does not attempt to refute these experimentally based quantitative arguments. Consequently, it's hard to see how my decision not to make more of these enigmatic small shelly fossils in any way undermines the main arguments of *Darwin's Doubt*.

Notes

1. Specifically, Marshall charges: "Meyer completely omits mention of the Early Cambrian small shelly fossils and misunderstands the nuances of molecular phylogenetics, both of which cause him to exaggerate the apparent suddenness of the Cambrian explosion." Charles R. Marshall, "When Prior Belief Trumps Scholarship," *Science* 341 (September 20, 2013): 1344.

2. Stephen C. Meyer, *Darwin's Doubt: The Explosive Origin of Animal Life and the Case for Intelligent Design* (New York: HarperOne 2013), on 425 in the hardback and 460 in the paperback (footnote 39 for Chapter 3, "The *Not* Missing Fossils").

3. John Farrell, "How Nature Works," *National Review*, September 2, 2013, https://www.nationalreview.com/nrd/articles/355862/how-nature-works; Stephen C. Meyer, "Further Debate on the *Origin of Species*, *National Review* September 30, 2013, https://www.nationalreview.com/nrd/articles/358310/letters.

4. See Figure 1, Charles R. Marshall, "Explaining the Cambrian 'Explosion' of Animals," *Annual Reviews of Earth and Planetary Sciences* 34 (2006): 355–384.

5. Charles R. Marshall, "Explaining the Cambrian 'Explosion.'"

6. See *Darwin's Doubt*, Chapter 10.

7. See *Darwin's Doubt*, Chapter 12.

IV.
BIOLOGIST:
MARTIN POENIE

There is no debate about evolution.

PHYSICIST AND THEISTIC EVOLUTIONIST KARL GIBERSON
Karl Giberson, "The Crazy Way Creationists Try To
Explain Human Tails Without Evolution," *The Daily
Beast*, June 1, 2014, http://www.thedailybeast.com/
articles/2014/06/01/the-crazy-way-creationists-try-
to-explain-human-tails-without-evolution.html.

15.

ANSWERING OBJECTIONS FROM

MARTIN POENIE

Douglas Axe

IN JUNE 2013, UNIVERSITY OF PITTSBURGH PHYSICIST DAVID SNOKE posted a favorable review[1] of Stephen Meyer's book, *Darwin's Doubt*, on the website of the Christian Scientific Society. Someone writing under the name "gandaulf" thought it was *too* favorable, judging by the series of critical comments he posted in response.

Although most anonymous comments don't merit a reply, I knew from multiple credible sources that this gandaulf is a serious scientist: molecular cell biologist Martin Poenie from University of Texas at Austin. I identify him here with his permission. Since some of Poenie's criticisms touch on my work, I'll offer my perspective.

Poenie's first critical comment questions Meyer's basis for thinking that the Cambrian explosion must have involved the origin of many new protein folds. According to Poenie (gandaulf), "The argument that many new folds are needed at the Cambrian explosion is without foundation."[2]

I suppose we could approach this topic by putting on either of two hats: the hat of an engineer (someone who designs things) or the hat of a *reverse* engineer (someone who dissects things to gain some understanding of how they were designed). But considering how far human technology is from designing anything like life, it would be presumptuous for any of us to wear the engineer's hat here. The role of the reverse engineer is much humbler, and much more appropriate.

Poenie may be thinking that Meyer made the mistake of putting the engineer's hat on, speaking about what is needed to build an animal as if he knows how to build one. But any reasonably charitable reading of Meyer would suggest that in raising the question about the requirements for building complex animals, he was approaching the question retrospectively in the manner of someone attempting to reverse engineer these systems. So let's assume that he wrote from the perspective of a reverse engineer, not claiming to have mastered the art of making new animals, but rather recounting some of the things science has established after considerable experience in the study of cells and the dissection of animals, both genetically and anatomically, about what the evolutionary process would have needed to generate in order to build a novel form of animal life.

One well-established fact is that individual species carry lots of genes that, so far as we can tell, are unique to their kind. If you search Google Scholar for the term *orphan genes*, you'll get over a hundred thousand results. According to a recent paper, "Orphan genes are defined as genes that lack detectable similarity to genes in other species and therefore no clear signals of common descent (i.e., homology) can be inferred."[3] The term is also sometimes applied to genes that are restricted to groups at a higher level than species, the key point being that many, many genes are specific to particular taxonomic types. In fact, a whopping *majority* of the full catalog of gene types identified by genome sequencing projects appears to be restricted in this way. As that recent paper put it, "only a small set of genes seems to be universal across kingdoms, whereas the phylogenetic distribution of all other genes is restricted at different levels."

Now, since each gene carries the sequence instructions for making a protein, it seems likely that orphan genes tend to encode orphan proteins—proteins that are substantially distinct from any found in other kinds of organisms. And if so, it also seems likely that many of these orphan proteins have distinct structures, or *folds*, as they are known.

Again, we could criticize this claim on the grounds that no one presently knows how to design new protein folds with any proficiency, but this is pointless because reverse engineering has shown that the inference is correct. Proteins with no detectable similarity to any protein of known structure have been found to have unique fold structures in about half of the cases examined.[4] Considering that orphan genes typically account for 10 percent to 30 percent of the genes in each sequenced genome, and that multicellular animals have about ten thousand or more genes, this means we can expect to find many dedicated protein folds in each specific kind of animal, right down to the level of species.[5]

So while the passage of half a billion years prevents us from actually examining the proteins that were used within the cells that made up the animals that appeared in the Cambrian explosion, the diversity and number of these animal forms leads us to believe that there must have been a corresponding explosion of protein forms. This certainly follows from the facts as we now see them, so Poenie's assertion is misinformed.

To me his assertion also seems a bit disingenuous, in that Poenie appears to be trying to dismiss a critical problem without answering it. Protein folds are a biological reality, presently catalogued by the thousands with more being added all the time. So any theory of biological origins that can't explain the origin of protein folds is in trouble. Period.

Drawing on a wide body of evidence, I've argued in detail that Darwinian evolution *is* in trouble for precisely this reason.[6] Failure to explain protein folds certainly isn't the only trouble plaguing Darwinism, but it *is* major trouble of a particularly stark kind that only gets worse as the science progresses. Poenie ought to grapple with this instead of trying to sweep it under the rug.

Notes

1. David Snoke, "Review of Steve Meyer's New Book, 'Darwin's Doubt,'" *The Christian Scientific Society*, June 21, 2013, http://www.christianscientific.org/review-of-steve-meyers-new-book-darwins-doubt/.

2. "gandaulf," *The Christian Scientific Society*, June 23, 2013, http://www.christianscientific.org/review-of-steve-meyers-new-book-darwins-doubt/#comment-224.

3. Lothar Wisler et al., "Mechanisms and dynamics of orphan gene emergence in insect genomes," *Genome Biology and Evolution* (2013), http://gbe.oxfordjournals.org/content/early/2013/01/24/gbe.evt009.full.pdf+html.

4. Sung-Hou Kim et al., "Structural genomics of minimal organisms and protein fold spaces," *Journal of Structural and Functional Genomics* 6 (2005): 63–70, http://compbio.berkeley.edu/people/brenner/pubs/kim-2005-jsfg-minimal.pdf.

5. Lothar Wissler, "Orphan Gene Emergence."

6. Douglas D. Axe, "The Case Against a Darwinian Origin of Protein Folds, *BIO-Complexity* 2010, no. 1 (2010), 1–12, http://bio-complexity.org/ojs/index.php/main/article/view/BIO-C.2010.1.

16.

More on Objections from Martin Poenie

Douglas Axe

Tʜɪs ɪs ᴍʏ sᴇᴄᴏɴᴅ ʀᴇᴘʟʏ ᴛᴏ ᴄᴏᴍᴍᴇɴᴛs ᴛʜᴀᴛ Uɴɪᴠᴇʀsɪᴛʏ ᴏꜰ Texas at Austin biologist Martin Poenie posted at the web site of the Christian Scientific Society. Writing under the name gandaulf, Poenie critiqued arguments that Stephen Meyer made about proteins in *Darwin's Doubt*. See the preceding chapter for more background.

Poenie goes on:

To add a bit more, I think Meyer could be so much more compelling to examine some wider data than just what Doug Axe says. I do not know Doug and have nothing against him. Furthermore, I could not care less if the Darwinian paradigm falls apart on the Cambrian explosion. But there is what is known as the Ig superfamily of proteins which contain, as the name suggests, the Ig fold. Members of the Ig superfamily are involved in homotypic adhesion (the foundation for making tissues), receptors, signaling proteins (tyrosine kinases) and of course antibodies and T cell receptors. Now here are two points that are interesting. First, one fold is used for many different and relevant functions, and one in particular that lies at the heart of the Cambrian explosion—mulicellularity—which involves homotypic adhesion. Secondly, what is the sequence variability of the Ig fold in members of the superfamily. If it is as constrained as Meyer portends, then we should see it in the sequence data.[1]

This continues the thought of his first comment, which is that Meyer has no basis to think that the striking variety of animal forms that appeared in the Cambrian explosion would have required new protein folds. The gist of my response was that the distribution of unique genes and proteins (orphans, as they are often called) among extant animal kinds shows that each of the different kinds, right down to the level of species, carries many genes and proteins that are unique—found nowhere else. So even if it is conceivable that animal life in all its diversity could have been formed without lots of new protein folds, that idea is purely hypothetical in that life as we see it doesn't seem to have been produced that way. Economizing on protein folds doesn't seem to have been a priority.

In defense of his assertion that new protein folds are not needed for new animal forms, Poenie points out in the above comment that one fold can perform many functions, citing the immunoglobulin fold as a key example. The reasoning here is that once life has a basic set of protein folds, it should be easy for evolution to produce any number of new protein functions by reusing those folds.

Aside from the fact that life doesn't seem to have limited itself to a common set of basic folds (which was my first point) there is another problem with Poenie's reasoning, one that is prevalent in evolutionary thinking. The root of the problem lies in a curious difference between the way biologists think of their science and the way chemists or physicists think of theirs. Biologists, unlike the others, tend to think they are doing science when they name things.

I remember Glenn Seaborg from my undergraduate days at Berkeley, a man who had the honor of naming several elements in the periodic table, and who had the even higher honor of an element being named after him (seaborgium, atomic number 106). But in each of these instances, the science was done well before the naming, and no one thinks the naming of an element establishes anything about its properties.

In biology, on the other hand, names are loaded with interpretations to the point where the boundaries between nomenclature and scientific facts are badly blurred. Yuri Lazebnik expressed this very well in a hilarious essay on "the common fundamental flaw of how biologists approach problems," which he titled "Can a Biologist Fix a Radio?"[2]

Like most biologists, Poenie takes a grouping convention, namely the grouping of proteins into sets called *families* or *superfamilies* or *folds*, to be significant in itself. My question is, how significant can a convention of that kind really be? If you knew nothing about two protein domains other than that they are said by convention to have the same fold, what would you be able to infer from that?

For example, these two protein domains, colored from red to blue along their chains, are classified as having the immunoglobulin fold. On the left is the N-terminal domain from sweet potato purple acid phosphatase (PDB accession 1xzw), as classified by the Structural Classification of Proteins.[3] On the right is a domain from the extracellular region of human tissue factor (PDB accession 2c4f), from the same classification source.

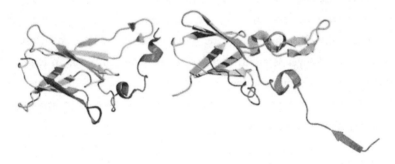

Figure 16-1. Illustration: Douglas Axe/Biologic Institute

Both of them are portions of larger structures that perform very different functions. These portions have both similarities and differences. In fact, the differences are large enough that, apart from some of the arrow shapes (beta strands) on the left sides of the pictures, there is no clear correspondence of parts even at the level of gross structure, much less the finer level of the genetically encoded amino acids from which the structures are made (not shown).

I'm not suggesting that there is no basis for classifying these domains as belonging to a fold group, or that protein classification schemes are unimportant. Indeed, classification is a key step toward bringing conceptual order to what would otherwise be a bewildering assortment of individual protein structures and functions. What I *am* saying is that the real scientific question of where the many distinct protein structures and functions came from is not reduced or collapsed simply by placing them into groups. If it were, then we could all but eliminate the problem by grouping them under the single heading: *proteins.*

As it is, the fact that the above domains are conventionally grouped together merely provides a convenient way of grouping the key questions, all of which remain unanswered. Can either of these replace the other without loss of function? Do they have parallel histories, or common histories? Did they evolve in Darwinian fashion from a common starting point? None of these questions is eliminated by the convention of referring to these domains as examples of the immunoglobulin fold. Indeed, the fact that their similarities are lost within the whole protein structures that contain them (below) makes it hard to guess whether those similarities have anything to do with the answers to these hard questions. Meyer is right, then, to refer to protein folds in explaining the problems they pose for Darwinian evolution, and Poenie is wrong to think that the existence of these groupings somehow eliminates the problems that Meyer raises.

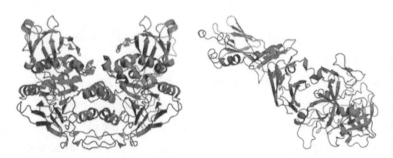

Figure 16-2. Illustration: Douglas Axe/Biologic Institute

In the end, the only way to find out whether structural similarity has any connection to material equivalence or evolutionary relatedness is to perform experiments. My colleague Ann Gauger and I have been doing this for some time now to address the specific question of whether similarity implies that evolutionary transitions are feasible. As skeptics, we decided to look at a pair of enzymes (proteins that do chemistry) with much more striking structural similarity than biologists require in order to infer evolutionary relatedness.

We chose to study this pair of enzymes, called Kbl and BioF:

Figure 16-3. Illustration: Douglas Axe/Biologic Institute

The structural similarity here is so clear that part-for-part correspondence is completely unambiguous, right down to the level of the individual amino acids from which the chains are built (not shown here). Despite the fact that they are structural twins in this sense, the two enzymes catalyze different chemical reactions with no functional overlap.

We asked whether Darwinian evolution is capable of transforming one of these enzymes to perform the function of the other. Think of this as a much more modest version of the kind of functional transition that Poenie thinks we can safely infer from much less striking similarities among proteins classified as having the immunoglobulin fold. If the logical leap from vague similarity to evolutionary relatedness is really justified, as biologists commonly assume, then transitions between enzymes with striking similarity ought to be a snap. Conversely, if transitions between enzymes with striking similarity are found *not* to be a snap, then biologists ought to start questioning those logical leaps.

As Ann and I reported, the transition from Kbl function to BioF function appears to be an evolutionary impossibility.[4] Furthermore, we haven't seen a convincing case that *any* evolutionary transition from one enzyme function to a genuinely different one *is* feasible. Even if compelling examples are eventually found, the general difficulty of functional transitions is now well-established, and that in itself makes the uncritical inference of evolutionary relatedness from similarity alone bad science. Of course, when you consider the central role this uncritical inference plays in evolutionary reasoning, you'll understand why evolutionary biologists are loath to rethink it.

Like most biologists, Martin Poenie thinks that fold similarity proves evolutionary relatedness, and because of this he thinks that sequence variability among proteins classified as having the immunoglobulin fold should be a good indicator of constraints. I've spent many years examining this kind of reasoning, and I've found it to be unsound. If Poenie is willing to examine the evidence, he might find himself agreeing with me.

Notes

1. Martin Poenie (as gandaulf), comment on "Review of Steve Meyer's New Book, *Darwin's Doubt*," *The Christian Scientific Society*, June 23, 2013, http://www.christianscientific.org/review-of-steve-meyers-new-book-darwins-doubt/#comment-225.

2. Y. Lazebnik, "Can a Biologist Fix a Radio?—or, What I Learned while Studying Apoptosis," *Biochemistry* (Moscow) 69, no. 12 (2002): 1403–06, http://protein.bio.msu.ru/biokhimiya/contents/v69/pdf/bcm_1403.pdf.

3. See http://scop.mrc-lmb.cam.ac.uk/scop/index.html.

4. Ann K. Gauger and Douglas D. Axe, "The Evolutionary Accessibility of New Enzymes Functions: A Case Study from the Biotin Pathway," *BIO-Complexity* 2011, no. 1 (2011): 1–17, http://bio-complexity.org/ojs/index.php/main/article/view/BIO-C.2011.1.

17.

ORPHAN GENES:

A GUIDE FOR THE PERPLEXED

Ann Gauger

Editor's note: Though Martin Poenie contributed a reply to Douglas Axe, Dr. Poenie declined to give permission to include his post in this book. Logically, his article should have preceded this one. It may still be read at Evolution News & Views.[1]

A S READERS WILL KNOW WHO HAVE FOLLOWED THE EXCHANGE between Martin Poenie and Doug Axe, originating in critical comments by Dr. Poenie about Stephen Meyer's argument in *Darwin's Doubt*, we have a dispute going on between scientists over how to interpret experiments about proteins. The dispute has to do with whether or not neo-Darwinian evolution, or any other form of unguided evolution for that matter, has the creative power attributed to it by many scientists.

Living things depend on a myriad of proteins to carry out cellular functions. These proteins are not globs of unstructured stuff. To carry out their function in the cell, most proteins have to fold into particular three-dimensional shapes. Interestingly, most proteins can be arranged into groups based on the similarity of their structures, or folds, as scientists call them. There are several thousand distinct folds now known among proteins whose structure has been determined.

Here's the surprising thing. As scientists sequence more genomes from different organisms, they are discovering that roughly 10–20 percent of each genome's protein-coding sequence is *new*, that is, *unlike any other known protein-coding sequence*. This was a one of the biggest surprises to come out of the whole genome-sequencing project, though by no means the biggest.

Why? The working assumption had been that, given common descent and the fact that most housekeeping genes are shared among living things, and the assumption hitherto that evolution occurs by incremental small changes, orphan genes (protein-coding sequences without known protein-coding antecedents) should be rare if not non-existent.

At this point it is necessary to explain a little about how such orphan sequences come to be identified. When such DNA is copied into RNA, substituting U for T, the RNA is then interpreted by the protein-making machinery using the following code: AUG tells the protein-making machinery, "Start here," and UAA, UAG, and UGA say, "Stop here." Hence the names "start" and "stop" codons. Just statistically speaking, stop codons should be relatively common in a random DNA sequence. In DNA that does not code for protein, roughly 1 in 20 triplet sequences will be either TAA, TAG, and TGA. Therefore, stretches of DNA that have a start codon and no in-frame stop codons for at least 100 nucleotides or more are called open reading frames, or ORFs (the length chosen depends on assumptions made about what constitutes a minimum length for protein function), and on that basis are identified as possible protein-coding genes (this is the case in bacteria—in eukaryotes it's more complex).

Orphan genes (sometimes called ORFan genes in bacteria) are those *open reading frames that lack identifiable sequence similarity to other protein-coding genes*. Lack of similarity is hard to prove, given the size of the genomic universe. Methods vary from researcher to researcher, so each study needs to be evaluated carefully. There is also always the possibility that any given ORF has no function. No doubt some orphan genes will prove to be artifacts of incomplete evidence (see below). But orphan genes are a reality, nonetheless, based on numerous and substantial studies.

The existence and prevalence of orphan genes raises a number of significant questions.

1. *Do orphan genes encode functional proteins?* In many cases there is evidence to suggest that they do. Some are highly conserved, even essential for viability to the organism from which they come, implying they are functional. Some are known to be involved in important species-specific or group-specific functions.

2. *Will similar sequences be found in other genomes, as we obtain more data?* This could be the case if genes classified as orphans are simply the result of our having sampled too little of worldwide genomic diversity. As more genomes are sequenced, we may find that orphan genes are not alone. Orphan genes could be examples of once common genes now lost in most other species, or they could be far voyagers, come from other life forms and integrated into new contexts (this is especially possible among bacteria). This is unlikely to be the case for all orphan genes, however, because we keep discovering new ones as we sequence more genomes.

3. *Will orphan proteins show structural similarity, if not sequence similarity, to known proteins?* This would suggest that orphan genes started out similar, but have lost their similarity because of rapid adaptive evolution or, alternatively, long-term neutral evolution. The current answer would seem to suggest that at least some orphan genes have no known structural similarity and are therefore unrelated to other known proteins. It is too soon to say whether that will always be the case.

4. *The fact that such surprising species- or clade-specific proteins exist raises interesting questions about where orphans come from.* Some might have come from gene duplication (duplication of coding DNA) followed by rapid adaptive evolution (see #3 above). If that is the case we should see traces left behind in the orphan

protein's three-dimensional structure. Another possible mechanism might be recruitment from non-coding DNA by a combination of mechanisms, including insertion of transposable elements. This is possible, but it would require that the insertion or other mechanism(s) be lucky events in order to produce a stable, functional protein, that is, one that is of use to the organism. Exactly how lucky such an event might be is one of the issues we are debating.

5.   Then there is the elephant in the room that evolutionary biologists don't want to acknowledge. *Perhaps we see so many species- and group-specific orphan genes because they are uniquely designed for species- and group-specific functions.* Certainly, unique design runs contrary to the expectation of common descent.

Exciting times! Much more work has to be done before we can determine which of the above possibilities are true. It may well be that all of them are true, at least sometimes, though I am sure Dr. Poenie would rule out #5. If common descent is true, the apparent rate of generation of new proteins is astonishing by anyone's expectation. What now needs to be determined is whether or not naturalistic processes known to be operating are *actually capable* of generating so many new proteins.

Notes

1. Martin Poenie, "Douglas Axe, Protein Evolution, and *Darwin's Doubt*: A Reply," *Evolution News & Views*, July 23, 2013, http://www.evolutionnews.org/2013/07/douglas_axe_pro074781.html.

18.

PROTEIN EVOLUTION:

A GUIDE FOR THE PERPLEXED

Ann Gauger

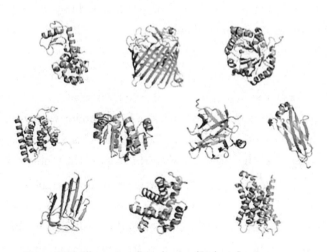

Figure 18-1. Illustration: Douglas Axe/Biologic Institute

PROTEINS COME IN MANY SHAPES AND SIZES, AS YOU CAN SEE IN the illustration above taken from Doug Axe's paper, "The Case Against a Darwinian Origin of Protein Folds."[1] Some can perform their functions as a single folded "domain," a coherent, stably folded unit of protein structure. Others are composed of multiple linked domains, or even separate folded chains that must come together to form the functional unit that is useful to the cell.

How proteins manage to fold into the correct shape is an area of active study that others have discussed recently here. What I want to address is why the problem of protein evolution is such a big deal, and what the disagreement with Martin Poenie is about. Dr. Poenie, a University of Texas biologist, critiqued Stephen Meyer's book, *Darwin's Doubt*, leading to an exchange between Poenie and Doug Axe.

As Steve Meyer explained in his book, the problem is that the number of possible protein sequences that *could* exist is *very large*, occupying a very large potential sequence space, but the number of proteins that *do* exist is much smaller, and they are widely scattered across sequence space (perhaps—in fact, that is one of the things being debated). The potential space is so large that a purely random search for *rare* functional proteins would spectacularly fail. So unless functional sequences are easy to find (very common), and/or are clustered together (easily reachable from one functional island to another), explaining current protein diversity *without design* is impossible.

Our interlocutors are keenly aware of the problem. To solve it, some propose that the first proteins were composed of just a few kinds of amino acids. Others propose that the first proteins were very small, or that they were very non-specific ("promiscuous" is the word used in the literature). These hypotheses reduce the scope of the problem somewhat.

Others like to suggest that, having somehow stumbled on one or a few successful folds, evolution was able to bootstrap its way forward by a combination of gene duplication and cooption of the duplicates to new functions, or by recombining existing proteins into new functional forms. There is now also the suggestion that completely new proteins can sometimes be generated by the random insertion of mobile genetic elements into non-coding DNA, though this was thought to be very unlikely just a few years ago.

The problem with these scenarios is that they require proteins to be remarkably tolerant of sequence rearrangements and insertions, or remarkably easy to shift to new functions, or remarkably different at the dawn of life from what they are now—large, complex, macromolecules tailored for specific functions.

Given that no one knows how the chemistry of life could be carried out by a handful of non-specific proteins, most people opt for explanations having to do with recruitment of duplicate proteins to new functions by point mutation or rearrangement, or the *de novo* creation of new protein-coding genes from previously non-coding DNA.

Here is where it gets interesting. If you examine the arguments made by Martin Poenie,[2] they appear to be mutually contradictory:

1. Proteins can be recruited to new functions, but only if you start with the right ancestral form. This is because multiple chemical interactions are required to give a protein its shape and catalytic activity; proteins that differ too much in sequence can have completely different networks of interactions, even if they share the same shape. This means that there may be no stepwise path to convert one to the other. Proteins are finicky things.

2. Introducing a few mutations can so disrupt an enzyme that its delicate catalytic activity is destroyed, making it impossible to recruit enzymes to new functions that are not already very similar in sequence, and/or already share some level of function. Once again, proteins are finicky things.

Or:

3. Proteins are robust, and can easily be improved by recombining them. In this way, new combinations of mutations can be produced in one fell swoop, sidestepping potential non-functional combinations.

4. You can create a new functional gene by inserting whole chunks of DNA into prior non-coding DNA, that by blind luck, not design, is transcribed and translated into a new functional protein (something the organism actually benefits from).

So which is it? Hard or easy? Remember, for evolution to work, proteins need to be remarkably tolerant of sequence rearrangements, or remarkably easy to shift to new functions by amino acid substitutions, or functional sequences need to be quite common.

It would appear that #1 and #2 say the evolution of proteins is hard to explain. We agree. If it were easy to shift proteins to new functions, then something like the transition between *Kbl* and *BioF* should have been possible.[3] It proved not to be possible. But claiming that the reason we failed was because we didn't start with exactly the right ancestral form makes life's history one long providential journey, or the product of an incredibly lucky series of accidents. Luck is a shaky foundation upon which to build the story of evolution, leaving design as the alternative.

The paper by Romero and Arnold[4] that Poenie cites in support of claim #3 is based on experiments that have nothing to do with the problem of Darwinian evolution, and everything to do with genetic engineering. The researchers sought to optimize already existing enzymes by recombining family members that already had the same structure and function, but different amino acid sequences. To ensure the greatest likelihood of success, the experimenters used an algorithm called SCHEMA to carefully choose recombination breakpoints at positions most likely to generate proteins that work. So this experiment says nothing about what *random* recombination can do, or its ability to generate *new* function. In fact the authors of this study clearly state that random recombination fails miserably by comparison. Thus inserting random sequences into old proteins is unlikely to be a source of new functional proteins.

Claim #4 is based on the fact that novel protein-coding sequences (orphan genes) exist in our own genomes and elsewhere. But just to point out what should be obvious, the fact *that something exists* does not explain *how it came to exist*. Unless it can be shown that insertion of elements into a random non-coding sequence *in an unguided fashion* really does produce functional proteins, then we can talk about it all we want, but we still don't know how the orphan genes got there.

Finally Doug Axe and others have had something to say about the rarity of functional folds. His 2004 paper, "Estimating the Prevalence of Protein Sequences Adopting Functional Enzyme Folds," came up with a shockingly small number. From the abstract:

> Starting with a weakly functional sequence carrying [the pattern of hydropathic constraints along chains that form the beta-lactamase domain fold], clusters of ten side-chains within the fold are replaced randomly... and tested for function. The prevalence of low-level function in four such experiments indicates that roughly one in 10^{64} signature-consistent sequences [having the same hydropathic constraints] forms a working domain.[5]

That number is based on experiment with real enzymes, not *in silico* mock-ups.

So which is it? From the above claims and rebuttals, it is reasonable to conclude that proteins are not tolerant of sequence rearrangements or recombination with non-coding DNA. Neither are they easy to shift to new functions by amino acid substitutions. Lastly, functional sequences are quite rare. From these statements, it would appear that the unguided evolution of proteins is hard, very hard. In fact, all the evidence points in the opposite direction. And that leaves us with the conclusion that explaining current protein diversity without design is very hard indeed.

Notes

1. Douglas D. Axe, "The Case Against a Darwinian Origin of Protein Folds," *BIO-Complexity* 2010, no. 1 (2010), 1–12, http://bio-complexity.org/ojs/index.php/main/article/view/BIO-C.2010.1.

2. Martin Poenie, "Review of Axe's work by Martin Poenie," *The Christian Scientific Society*, July 22, 2013, http://www.christianscientific.org/review-of-axes-work-by-martin-poenie.

3. See Ann K. Gauger and Douglas D. Axe, "The Evolutionary Accessibility of New Enzyme Functions: A Case Study from the Biotin Pathway, *BIO-Complexity* 2011, no. 1 (2011), 1–17, http://bio-complexity.org/ojs/index.php/main/article/view/BIO-C.2011.1.

4. Philip A. Romero and Frances H. Arnold, "Random Field Model Reveals Structure of the Protein Recombinational Landscape," *PLOS Computational Biology*, October 4, 2012, http://dx.doi.org/10.1371/journal.pcbi.1002713.

5. Douglas D. Axe, "Estimating the Prevalence of Protein Sequences Adopting Functional Enzyme Folds," *Journal of Molecular Biology* 341 no. 5 (August 27, 2004), 1295–1315, http://dx.doi.org/10.1016/j.jmb.2004.06.058.

19.

SHOW ME:

A CHALLENGE FOR

MARTIN POENIE

Douglas Axe

I APPRECIATED UNIVERSITY OF TEXAS CELL BIOLOGIST MARTIN POE-
nie's taking the time to reply to what I said in Chapters 15 and 16.[1]

Poenie repeats a complaint that many others have made about the
study of BioF that I did with Ann Gauger. The complaint is that we
examined the difficulty of a non-historical functional transition instead
of attempting to reconstruct evolutionary history, and this supposedly
makes our negative result irrelevant.

Ann and I anticipated this criticism and explained what's wrong
with it in our paper,[2] and I've reiterated our point at least two times since
then.[3] The problem, once again, is that biologists like Poenie want to be
free to appeal to evolutionary processes for explaining past events without shouldering any responsibility for demonstrating that these processes actually work in the present. That clearly isn't valid. Unless we want
to rewrite the rules of science, we have to assume that what doesn't work
didn't work.

It isn't valid to think that evolution *did* create new enzymes if it hasn't been demonstrated that it *can* create new enzymes. And if Poenie really thinks this has been done, then I'd like to present him with an opportunity to prove it. He says, "Recombination can do all the things that Axe thinks are impossible." *Can it really? Please show me, Martin!*

I'll send you a strain of *E. coli* that lacks the *Kbl* and *BioF* gene, and you show me how recombination, or any other natural process operating in that strain, can create a new gene that does the job of *BioF* within a few billion years. You wouldn't have to run a billion-year experiment to do this. You would simply have to characterize an actual process in your lab, including the constraints within which it operates, and then do the math to show what it would do in a realistic population over an evolutionary timeframe.

That's exactly what Ann and I did. And I suspect that if you, Martin, were to take this challenge seriously, *then* our approach would start to make sense to you. And if you find a different approach that really does show how a working replacement for *Kbl* and *BioF* can evolve, I assure you that you will have my rapt attention and my full respect. After all, that's what real science deserves.

But evolutionary groupthink? Hmmm... not so much.

Notes

1. Martin Poenie, "Douglas Axe, Protein Evolution, and Darwin's Doubt: A Reply," *Evolution News & Views*, July 24, 2013, http://www.evolutionnews.org/2013/07/douglas_axe_pro074781.html.

2. Ann K. Gauger and Douglas Axe, "The Evolutionary Accessibility of New Enzymes Functions: A Case Study from the Biotin Pathway," *BIO-Complexity* 2011, no. 1 (2011), 1–17, http://bio-complexity.org/ojs/index.php/main/article/view/BIO-C.2011.1.

3. See: Doug Axe, "When Theory and Experiment Collide," *Biologic Institute*, April 16, 2011, http://www.biologicinstitute.org/post/18022460402/when-theory-and-experiment-collide; Doug Axe, "Are We Reaching a Consensus that Evolution is Past its Prime?," *Biologic Institute*, October 25, 2012, http://www.biologicinstitute.org/post/34190339725/are-we-reaching-a-consensus-that-evolution-is-past-its.

V.

MISCELLANEOUS CHALLENGES

*There is no controversy in the mainstream
scientific community about either
the fact of evolution or the major
aspects of evolutionary theory.*

BARBARA FORREST, PHILOSOPHER,
SOUTHEASTERN LOUISIANA UNIVERSITY

Barbara Forrest, "Understanding the Intelligent Design
Creationist Movement: Its True Nature and Goals,"
The Center for Inquiry and Office of Public Policy (May
2007, amended July 2007), http://www.centerforinquiry.
net/uploads/attachments/intelligent-design.pdf.

20.

BACKHANDED COMPLIMENTS FROM

THE NEW YORKER

David Klinghoffer

SOME OF THE MOST INTERESTING, POSITIVE, AND EXCITING COVER-age of Thomas Nagel's book *Mind and Cosmos: Why the Materialist Neo-Darwinian Conception of Nature Is Almost Certainly False* came from what you might have thought were unlikely sources: liberal, secular venues like *The New Republic* and *The New York Review of Books*.[1] Atheist philosopher Dr. Nagel had, of course, aroused fury from the materialist posse with his praise for theorists of intelligent design, notably Stephen Meyer.

But the welcoming, almost relieved counter-response seemed to a mark a turning point. It looked like a long-standing condition of intellectual sclerosis was breaking apart and freeing up; suddenly, new ideas were getting through, receiving intelligent discussion in some very surprising places.

The blood of unimpeded debate was flowing again. The vow of silence was over.

Here's a hopeful sign that, in observing this, we were not mistaken. *The New Yorker* reviewed *Darwin's Doubt* and the result, though negative, was full of backhanded compliments. Ignore the snarkiness, and read between the lines. Linking to our announcement of Meyer's debut at #7 on the *New York Times* bestseller list,[2] Pulitzer Prize-winning science writer Gareth Cook concludes:

> [D]o not underestimate "Darwin's Doubt": it is a masterwork of pseudoscience. Meyer is a reasonably fluid writer who weaves anecdote and patient explanation. He skillfully deploys the trappings of science—the journals, the conferences, the Latinate terminology. He has a PhD from the University of Cambridge in the philosophy of science. He appears serious and, above all, reasonable. The Cambrian argument has been a part of creationism and its inheritors for many years, but Meyer's project is to canonize it, a task he completes with great skill. Those who feel a hunger for material evidence of God or who sense that science is a conspiracy against spiritual meaning will find the book a thrilling read. Which is to say, Meyer will find a large audience: he aims to start a conversation, or to at least keep one going, and he seems likely to succeed.
>
> The book's best, most honest moments come in the concluding chapter, in which Meyer travels to see the famous Burgess Shale in person. His son goes ahead on the trail but then suddenly freezes, stricken with vertigo after peering down the mountainside. Meyer likens his son's paralysis to modernity's despair at materialism, its shock at the prospect that the universe is utterly indifferent. Meyer writes frankly, saying that his quest is to give people back their sense of meaning and purpose. Here, at last, Meyer is not pretending to be a scientist.[3]

Sure, there are grounds to complain. Cook dismisses ID as "pseudoscience" in part because he lets grad student Nick Matzke's bogus response to Stephen Meyer's book, over at *Panda's Thumb*, carry much of the scientific burden of his own review for him. (See Casey Luskin's reply to Matzke, in Chapter 6 above.) Cook seems to have absorbed the National Center for Science Education's false narrative about the origins of ID, and I don't think he understands the argument that Meyer makes.

Cook says that with Darwinism having failed to explain the origin of the species (in Meyer's view), "The only alternative explanation, Meyer writes, is the involvement of an intelligent designer (read: God) who rushed along the story of life on Earth." For goodness' sake. No, that's not the "only alternative explanation": a source of designing intelligence, not necessarily God at all, is *the explanation that best fits the data*, as Meyer painstakingly argues. There's a big difference.

Cook writes:

> The problem for Meyer is that what has come to be called the Cambrian explosion was not, in fact, an explosion. It took place over tens of millions of years—far more time than, for example, it took humans and chimpanzees to go their separate ways. Decades of fossil discovery around the world, combined with new computer-aided analytical techniques, have given scientists a far more complete portrait of the tree of life than Darwin and Walcott had available, making connections between species that they could not see.[4]

The problem for Cook is that it's not the duration of the Cambrian event that's of interest. It's the fact that, whether in five, ten, or "tens" of million years, it brought into existence a bewildering variety of complex creatures that have no evident ancestry. *That* is the enigma, in the resolution of which "computer-aided analytical techniques" offer no aid since they point, unnervingly, to many conflicting Darwinian "trees of life."

Neither Cook nor Matzke reckons with Meyer's description and analysis of the competing post-Darwinian theories that are out there— forget about intelligent design! The scientific community is in the process of shrugging off a failed theory. What will replace it? That is news!

However, in perspective, these are minor grumbles about Cook's review. He seems to be a thoughtful guy—it would be interesting to sit in on a conversation between him and a few ID scientists. He would learn something and I think enjoy the experience.

What's important is the way the logjam against intelligent discussion of intelligent design in the mainstream media is finally unjamming. Guys like Jerry Coyne and Richard Dawkins (an increasingly eccentric figure) will continue to stonewall, refusing to evaluate or even acknowledge the arguments in *Darwin's Doubt* or other rigorous articulations of ID. Gentlemen like Nick Matzke will continue to search for typographical errors in our work, while creating scientific distractions to confuse the willing and the naïve.

That's all a sideshow. Real scientists and thoughtful, open-minded laymen are *paying attention right now* to a genuine and fascinating disputation about biological origins. The endorsements from scientists in relevant fields that *Darwin's Doubt* has already received is itself confirmation of that. I don't know how the debate will be resolved, if it ever will. But make no mistake: the debate is happening.

Notes

1. Alvin Plantinga, "Why Darwinist Materialism Is Wrong," *The New Republic*, November 16, 2012, http://www.newrepublic.com/article/books-and-arts/magazine/110189/why-darwinist-materialism-wrong; H. Allen Orr, "Awaiting a New Darwin," *The New York Review of Books*, February 7, 2013, http://www.nybooks.com/articles/archives/2013/feb/07/awaiting-new-darwin/.

2. David Klinghoffer, "Darwin's Doubt Debuts at #7 on *New York Times* Hardcover Nonfiction Bestseller List," *Evolution News & Views*, July 1, 2013, http://www.evolutionnews.org/2013/07/darwins_doubt_w073921.html.

3. Gareth Cook, "Doubting 'Darwin's Doubt,'" *The New Yorker*, July 2, 2013, http://www.newyorker.com/online/blogs/elements/2013/07/doubting-stephen-meyers-darwins-doubt.html.

4. Gareth Cook, "Doubting 'Darwin's Doubt.'"

21.

ABOUT AN ELLIPSIS:
JOHN FARRELL IN
NATIONAL REVIEW

David Klinghoffer

NATIONAL REVIEW DID ITSELF PROUD IN MAKING AMENDS FOR the silly John Farrell review of *Darwin's Doubt*. That's the one that climaxed with an examination of Stephen Meyer's placement of an ellipsis mark accompanied by vague intimations of scholarly malpractice.[1] Could such a momentous question as the one Meyer raises—whether biology gives evidence of design—ever be resolved in such a petty fashion, by a dispute over punctuation to be adjudicated by consulting a good copy editor? For sheer triviality this far exceeds even Nick Matzke and his "small shelly fossils."

In the September 30, 2013 issue, *NR* published a lengthy letter from Meyer, boxed and highlighted, prominently and politely demolishing Farrell. I think the placement sends a subtle message:

> As an avid reader of *National Review*, I'm honored that you would review my book *Darwin's Doubt*. Unfortunately, longtime intelligent-design critic John Farrell wildly misrepresents my argument and the current state of scientific evidence ("How Nature Works," September 2).

Contrary to what Mr. Farrell claims, *Darwin's Doubt* does not argue for intelligent design primarily based on the brevity of the Cambrian explosion, nor does it exaggerate that brevity. It affirms the widely accepted figure among Cambrian paleontologists of about 10 million years for the main pulse of morphological innovation in the Cambrian period that paleontologists typically designate as "the explosion." Nor does the book base its case for intelligent design upon "personal incredulity" about the creative power of materialistic evolutionary processes. Instead, it presents several evidentially based and mathematically rigorous arguments against the creative power of the mutation/natural-selection mechanism, none of which Farrell refutes.

The main argument of the book is not, as Farrell implies, a purely negative and, therefore, fallacious argument from ignorance. Instead, the book makes a positive case for intelligent design as an inference to the best explanation for the origin of the genetic (and other forms of) information necessary to produce the first animals. It does so based upon our experience-based knowledge of the power that intelligent agents have to produce digital and other forms of information. In formulating the argument as an inference to the best explanation, the book employs the same method of scientific reasoning that Darwin used in his *Origin of Species*.

Rather than engaging the actual arguments of the book, Farrell offers a spurious claim of out-of-context quotation, which has been amply refuted elsewhere by geologist Casey Luskin.[2] A genuine engagement with the debates currently taking place in evolutionary biology would have been far more interesting. Neo-Darwinism is fast going the way of other materialistic ideas such as Marxism and Freudianism, but readers of Farrell's review sadly were not able to learn why.

Stephen Meyer, Discovery Institute, Seattle, Wash.[3]

Farrell is then allowed a few lines to squeak briefly in his defense:

Stephen Meyer writes that his book "makes a positive case for intelligent design as an inference to the best explanation for the origin of the genetic (and other forms of) information necessary to produce the first animals."

But this presupposes something Dr. Meyer has never in fact demonstrated in a compelling fashion, either in this book or in his previous one: that new complex information cannot be generated by purely natural processes.

His inference to the best explanation—while one that some of his lay readers may be convinced of—to scientists is a copout. It is the job of scientists to find out how apparent design in nature can be explained by natural processes. The best explanation right now is Darwinian evolution.

That is the lamest rejoinder I've seen in a while. Farrell tries to save face by resorting to assertion. Demonstrating that "new complex information cannot be generated by purely natural processes" is exactly what Meyer does in *Signature in the Cell* and *Darwin's Doubt*, at length. Farrell never grappled with the actual evidence or arguments, and for this he was applauded by Jerry Coyne and the rest of the Darwin Defense Force—that would never applaud anything else in a conservative magazine like *NR*.

Farrell observes, "It is the job of scientists to find out how apparent design in nature can be explained by natural processes." But whether the design in nature is *real* or *apparent* is precisely the question that's up for debate. You don't settle it by slipping in an adjective, or quibbling over punctuation.

This guy is hopeless, but I knew that based upon uniform and repeated experience of him. Congratulations to *National Review* for setting things straight.

Notes

1. John Farrell, "How Nature Works," *National Review*, September 2, 2013, https://www.nationalreview.com/nrd/articles/355862/how-nature-works.

2. See Casey Luskin, "In *National Review*, John Farrell's Predictable and Misleading Review of Darwin's Doubt," *Evolution News & Views*, September 5, 2013, http://www.evolutionnews.org/2013/09/in_national_rev076261.html.

3. Stephen C. Meyer, "Further Debate on the *Origin of Species*," *National Review*, September 30, 2013, https://www.nationalreview.com/nrd/articles/358310/letters.

22.

DARWIN DEFENDERS LOVE

DONALD PROTHERO'S

AMAZON REVIEW

Casey Luskin

A FTER NICK MATZKE AT *PANDA'S THUMB* PUBLISHED A REVIEW of *Darwin's Doubt* that failed to preemptively knock down Stephen Meyer's thesis, the Internet's Darwin brigade was hoping for something better. So the folks at *Panda's Thumb* along with University of Toronto's Larry Moran and University of Chicago's Jerry Coyne were all excited when paleontologist Donald Prothero of the Natural History Museum of Los Angeles County posted an Amazon review of *Darwin's Doubt*.[1] Their readers eagerly voted up Prothero's post, artfully titled "Stephen Meyer's Fumbling Bumbling Cambrian Amateur Follies," as the "most helpful critical review."

According to Dr. Prothero, *Darwin's Doubt* is a mess of "fumbling," "bumbling," "distortions," and "blunders." The book is an "amateur" exercise, evidence of Meyer's "folly." It "butchers" the subject matter; was written by a "fool" who is "incompetent," guilty of "ignorance," is in "way over his head," and has a "completely false understanding of the subject." In case that's all a little too subtle for you, Prothero says Meyer argues "dishonestly" and promotes a "flat out lie," a "fundamental lie," and other "lies" to promote a "fairy tale."

Well, what justifies all the *ad hominem* invectives? Prothero's first complaint is that Meyer's PhD is in the history and philosophy of science which, according to Prothero, "give[s] him absolutely no background to talk about molecular evolution." That's a lame objection (it's called the genetic fallacy). Indeed, Meyer's undergraduate degree is in geology and physics, and he worked as a geophysicist for four years, giving him formal training on geology-related issues—the primary issues Prothero raises in his review. Prothero, however, has already undercut his own complaint, as he admitted *Evolution: What the Fossils Say and Why It Matters*:

> [Y]ou don't need a PhD to do good science, and not all people who have PhDs are good scientists either. As those of us who have gone through the ordeal know, a PhD only proves that you can survive a grueling test of endurance in doing research and writing a dissertation on a very narrow topic. It doesn't prove that you are smarter than anyone else or more qualified to render an opinion than anyone else.[2]

Prothero's review later complains that creationists "love to flaunt their PhD's on their book covers." I guess that means Meyer isn't a "creationist," since Prothero failed to notice that Meyer doesn't mention his PhD on the cover of *Darwin's Doubt*. (And isn't it a bit ironic that Prothero touts his own PhD in his bio over at Skepticblog?[3])

Prothero's second complaint is that "Almost every page of this book is riddled by errors of fact or interpretation that could only result from someone writing in a subject way over his head, abetted by the creationist tendency to pluck facts out of context and get their meaning completely backwards." Of course Prothero doesn't list examples from "almost every page," but at least this time he tries to give one. He claims "we now know that the 'explosion' now takes place over an 80 m.y. time framework." Perhaps Prothero didn't notice that Meyer *specifically discusses* Prothero's own view and *refutes* it in Chapter 3 of *Darwin's Doubt*. I answered the same argument in my response to Nick Matzke, which cited numerous articles from the mainstream technical literature stating that the Cambrian explosion took no more than 10 million years.[4]

Prothero goes on.

+ He objects that Meyer "dismisses the Ediacara fauna as not clearly related to living phyla," even though that's in fact the consensus view.[5]

+ He charges that Meyer "confuses crown-groups with stem-groups" (giving no examples), when in fact Meyer explains this distinction.[6]

+ He wrongly charges that ID is a "god of the gaps"[7] argument, one that invokes the "supernatural,"[8] when of course ID does no such thing,[9] and Meyer rebuts this charge decisively in Chapters 17 and 19 of *Signature in the Cell.*

+ He bizarrely misrepresents Meyer as saying Niles Eldredge and Stephen Jay Gould "are arguing that evolution doesn't occur," when Meyer said absolutely nothing of the kind.

Thus, a pattern in Prothero's review is that he puts words in Meyer's mouth, while failing to engage Meyer's actual arguments. This is seen again when Prothero writes:

Meyer deliberately and dishonestly distorts the story by implying that these soft-bodied animals appeared all at once, when he knows that this is an artifact of preservation. It's just an accident that there are no extraordinary soft-bodied faunas preserved before Chengjiang, so we simply have no fossils demonstrating their true first appearance, which occurred much earlier based on molecular evidence.[10]

Of course Meyer never says the Cambrian animals appeared "all at once." And did Prothero miss Chapter 5 of *Darwin's Doubt,* where Meyer discusses in great detail the "molecular evidence" mentioned by Prothero, meticulously critiques the molecular clock hypothesis, and clarifies why it doesn't account for the absence of evolutionary precursors in the Precambrian? Or what about Chapters 2 and 3, where Meyer explores the artifact hypothesis in much detail, and makes clear why many Cambrian experts feel it doesn't explain away the Cambrian explosion?

As Meyer observes, the Cambrian fossil record is full of soft-bodied organisms, making it difficult to argue that the lack of fossils from a particular group simply means they were too "soft-bodied" to have been preserved.[11] So it's not as if Meyer doesn't engage and discuss these issues; indeed, Meyer cites many authorities to show why these objections don't resolve the problem of the Cambrian explosion. Prothero neither engages with nor mentions any of these discussions.

Prothero asserts that the "rates of evolution during the 'Cambrian explosion' are typical of any adaptive radiation in life's history." Again, did he not read Section II of *Darwin's Doubt*, where Meyer argues that even if there were tens of millions of years available to evolve the Cambrian animals (as Prothero asserts), unguided evolutionary mechanisms still don't work fast enough to produce many of their complex features?

Prothero gives no indication that he has appreciated this section. Indeed his only specific objection is that Meyer supposedly "repeats many of the other classic creationist myths, all long debunked, including the *post hoc* argument from probability (you can't make the argument that something is unlikely after the fact)." This is a bizarre claim. Does Prothero not realize that many arguments for common ancestry are after-the-fact and probability-based—e.g., two similar gene sequences are unlikely to have arisen independently, and are thus said to have derived from a common ancestor?

My favorite part of Prothero's review comes when he says, "For a good account by real paleontologists who know what they're doing, see the excellent recent book by Valentine and Erwin, 2013, which gives an accurate view of the 'Cambrian diversification.'" Excellent indeed! Prothero is referring to Douglas Erwin and James Valentine's 2013 book, *The Cambrian Explosion*.

Let's look at what Erwin and Valentine have to say. Regarding the length of the Cambrian explosion, they write:

> [A] great variety and abundance of animal fossils appear in deposits dating from a geologically brief interval between about 530 to 520 Ma, early in the Cambrian period. During this time, nearly all the major living animal groups (phyla) that

have skeletons first appeared as fossils (at least one appeared earlier). Surprisingly, a number of those localities have yielded fossils that preserve details of complex organs at the tissue level, such as eyes, guts, and appendages. In addition, several groups that were entirely soft-bodied and thus could be preserved only under unusual circumstances also first appear in those faunas. Because many of those fossils represent complex groups such as vertebrates (the subgroup of the phylum Chordata to which humans belong) and arthropods, it seems likely that all or nearly all the major phylum-level groups of living animals, including many small soft-bodied groups that we do not actually find as fossils, had appeared by the end of the early Cambrian. **This geologically abrupt and spectacular record of early animal life is called the Cambrian explosion.**[12]

So it seems that unlike Prothero, Erwin and Valentine don't believe "the Cambrian explosion" took 80 million years, but rather that it took place during "a geologically brief interval between about 530 to 520 Ma."

Regarding the reality of the Cambrian explosion, Erwin and Valentine write:

> Taken at face value, the geologically abrupt appearance of Cambrian faunas with exceptional preservation suggested the possibility that they represented a singular burst of evolution, but the processes and mechanisms were elusive. Although there is truth to some of the objections, **they have not diminished the magnitude or importance of the explosion.... Several lines of evidence are consistent with the reality of the Cambrian explosion.**[13]

So it seems that, unlike Prothero, Valentine and Erwin don't believe the Cambrian explosion is merely an "artifact of preservation."

Regarding rates of evolution during the Cambrian explosion, Erwin and Valentine write:

> As geologists, we view this tension as a debate over the extent to which uniformitarian explanations can be applied to understand the Cambrian explosion. Uniformitarianism is often described as the concept, most forcefully advocated by Charles Lyell in his *Principles of Geology*, that "the present is the key to the past" (Lyell 1830). Lyell argued that study of geological processes operat-

ing today provides the most scientific approach to understanding past geological events. Uniformitarianism has two components. Methodological uniformitarianism is simply the uncontroversial assumption that scientific laws are invariant through time and space. This concept is so fundamental to all sciences that it generally goes unremarked. Lyell, though, also made a further claim about substantive uniformitarianism: that the rates and processes of geological change have been invariant through time (Gould 1965). Few of Lyell's contemporaries agreed with him (Rudwick 2008). **Today, geologists recognize that the rates of geological processes have varied considerably through the history of Earth and that many processes have operated in the past that may not be readily studied today.**

... One important concern has been whether the microevolutionary patterns commonly studied in modern organisms by evolutionary biologists are sufficient to understand and explain the events of the Cambrian or whether evolutionary theory needs to be expanded to include a more diverse set of macroevolutionary processes. **We strongly hold to the latter position.**[14]

Erwin and Valentine are skeptical that "uniformitarian explanations can be applied to understand the Cambrian explosion." Why? One reason could be that evolutionary mechanisms we observe in the present day operate at rates that are too slow to explain what took place in the Cambrian period. They are careful not to put it in such plain terms, but that is the essence of their argument. They do acknowledge that there was an "unusual period of evolutionary activity during the early and middle Cambrian"[15] and later expressly state:

> Because the Cambrian explosion involved a significant number of separate lineages, achieving remarkable morphological breadth over millions of years, the Cambrian explosion can be considered an adaptive radiation **only by stretching the term beyond all recognition... the scale of morphological divergence is wholly incommensurate with that seen in other adaptive radiations.**[16]

Unlike Prothero, Erwin and Valentine think the Cambrian explosion was a real event, took far less than 80 million years, and involved unique mechanisms that acted more rapidly and on a greater scale than other radiations. These beliefs directly contradict Prothero's core claims, but there's more.

Probably the most striking statement by Erwin and Valentine comes when they concede that we lack resolved evolutionary explanations for how the diversity of the Cambrian animals arose, and why these basic body plans haven't changed since that time:

> The patterns of disparity observed during the Cambrian pose **two unresolved questions.** First, what evolutionary process produced the gaps between the morphologies of major clades? Second, why have the morphological boundaries of these body plans remained relatively stable over the past half a billion years?[17]

Thus, when recently reviewing Erwin and Valentine's book, the journal *Science* stated: "The grand puzzle of the Cambrian explosion surely must rank as one of the most important outstanding mysteries in evolutionary biology."[18] Likewise, a 2009 paper in *BioEssays* stated, "elucidating the materialistic basis of the Cambrian explosion has become more elusive, not less, the more we know about the event itself."[19]

That pretty much nixes Prothero's confident, unbacked assertion that "scientists have explained most of the events of the Early Cambrian and find nothing out of the ordinary that defies scientific explanation."

What more is there to say? I wonder who in the community of Darwin-defenders will have a go at *Darwin's Doubt* next. The best of luck to them.

Notes

1. Donald Prothero, "Stephen Meyer's Fumbling, Bumbling Cambrian Follies," July 21, 2013, *Amazon.com*, http://www.amazon.com/review/R2HNOHERF138DU/.

2. Donald R. Prothero and Carl Dennis Buell, *Evolution: What the Fossils Say and Why It Matters* (New York: Columbia Univ. Press, 2007), 16.

3. "About Donald Prothero," *Skepticblog*, http://www.skepticblog.org/author/prothero/.

4. Casey Luskin, "How 'Sudden' Was the Cambrian Explosion?," *Evolution News & Views,* July 16, 2013, http://www.evolutionnews.org/2013/07/how_sudden_was_074511.html. This article is reprinted in the present book as Chapter 6.

5. See the section "The Significance of the Ediacaran," in Chapter 4 of *Darwin's Doubt*, 81–86 and endnotes.

6. Meyer, *Darwin's Doubt*, 419–20 (hardback) or 454–55 (paperback). See endnote 5 for Chapter 3, "Soft Bodies and Hard Facts."

7. Casey Luskin, "The Self-Refuting 'God of the Gaps' Critique," *Evolution News & Views*, October 18, 2012, http://www.evolutionnews.org/2012/10/the_self-refuti065411.html.

8. Casey Luskin, "ID Does Not Address Religious Claims About the Supernatural," *Discovery Institute*, September 8, 2008, http://www.discovery.org/a/7501.

9. Casey Luskin, "Principled (Not Rhetorical) Reasons Why Intelligent Designer Doesn't Identify the Designer," *Center for Science and Culture*, October 31, 2007, http://www.discovery.org/a/4306.

10. Prothero, "Fumbling, Bumbling."

11. Meyer, *Darwin's Doubt*, 62–64, in a section helpfully entitled "The Chengjiang Explosion."

12. Douglas H. Erwin and James W. Valentine, *The Cambrian Explosion: The Construction of Animal Biodiversity* (Greenwood Village, CO: Roberts & Co., 2013), 5. Emphasis added.

13. Ibid., 6. Emphasis added.

14. Ibid., 10. Emphasis added.

15. Ibid., 6. Emphasis added.

16. Ibid., 341. Emphasis added.

17. Ibid., 330. Emphasis added.

18. Christopher J. Lowe, "What Led to Metazoa's Big Bang?," *Science* 340 (June 7, 2013): 1170–1171, http://www.sciencemag.org/content/340/6137/1170.summary.

19. K. J. Peterson, M. R. Dietrich, and M. A. McPeek, "MicroRNAs and metazoan macroevolution: insights into canalization, complexity, and the Cambrian explosion," *BioEssays* 31, no. 7 (July 2009): 736–47, http://onlinelibrary.wiley.com/doi/10.1002/bies.200900033/abstract.

23.

DOES THE "GREAT
UNCONFORMITY" EXPLAIN
MISSING ANCESTORS?

Casey Luskin

URING THE Q&A AFTER STEPHEN MEYER'S LECTURE DEBUT
event for *Darwin's Doubt* at the Seattle Art Museum, a questioner
asked whether the "Great Unconformity" explains why the fossil record
does not bear fossils that are ancestors to the Cambrian animals.

An unconformity is an erosional surface representing a gap in the
geological record, where time is missing. We can recognize an unconformity when a geological layer sits directly on top of much older strata,
with a time-gap in between. The "Great Unconformity" is probably the
most famous such gap in the geological record, and is found in some
(though certainly not all) locations around the world. The exact time
span it represents is hard to define precisely (and probably varies significantly from location to location), though scientists generally suggest
that it extends from sometime in the Cambrian (perhaps as late as the
middle Cambrian) back hundreds of millions of years into the Precambrian world (perhaps as far back as 1.7 billion years ago). Thus, in some

places the "Great Unconformity" might represent over a billion years of missing time.

Stephen Meyer pointed out in response to the questioner that the "Great Unconformity" may be "worldwide" in the sense that it's found in many parts of the world. But that doesn't mean it's found *everywhere*. As Meyer explained, the Great Unconformity *cannot* be universal, otherwise we wouldn't have strata from the Ediacaran period, and we wouldn't know about Ediacaran-age fossils, such as the Precambrian sponge embryos Meyer talks about in *Darwin's Doubt*.

In the same way, the "Great Unconformity" in the Grand Canyon is thought to have erased significant parts of the Cambrian period. If this were the case everywhere, we would have no knowledge of the Cambrian explosion itself. This shows that while the "Great Unconformity" is a significant geological feature, it is not ubiquitous.

In response to the questioner, Meyer also pointed out that this is an unorthodox objection to the Cambrian explosion. I agree. Though I've read numerous articles struggling to explain away the Cambrian explosion, I'd never heard of anyone arguing that the alleged Precambrian ancestors to the Cambrian animals were missing because the strata containing their fossils had been completely eroded from the face of the earth in such a "Great Unconformity." But after hearing this objection, I did some research. I discovered a single place where this argument has been made: It was over 100 years ago by Charles Doolittle Walcott, the famous geologist who discovered the Burgess Shale. Apparently the argument was never adopted by subsequent geologists for good reasons.

In a 2006 paper in *Earth Science History*, Walcott biographer and former National Museum of Natural History paleobiologist Ellis L. Yochelson (now deceased) noted that Walcott gave up on finding biological explanations for the absence of Precambrian animal ancestors, and thus turned to geological ones. Yochelson quotes Walcott as follows:

> I have for the past eighteen years watched for geological and paleontological evidence that might aid in solving the problem of pre-Cambrian life. The great series... have been studied and searched for evidences of life until the conclusion has gradually

been forced upon me that on the North American Continent we have no known pre-Cambrian *marine* deposits containing traces of organic remains, and that the abrupt appearance of the Cambrian fauna results from geologic and not biotic consequences.[1]

Yochelson explains that Walcott dubbed this unconformity the "Lipalian interval," and used it as a geological explanation for the lack of Precambrian animal fossils:

> Whether Walcott was right or wrong in his interpretation, the Lipalian was a grand synthesis based on years of virtually fruitless searching for fossils in pre-Cambrian sedimentary deposits. There is an old saying that the absence of evidence is not evidence of absence. Somehow, that applies to the concept of the Lipalian and, equally, to the speculation as to what may have triggered that concept.[2]

This is just about the only source anywhere I can find where someone made this argument. Yochelson himself notes that Walcott's very notion of the "Lipalian interval" was never adopted or promoted by subsequent geologists. Much less have any recent geologists attempted to use it to explain away the lack of Precambrian fossils. Yochelson continues:

> The endeavor [Walcott's proposal of the Lipalian Interval] was for naught! So far as one can tell, the net result of Walcott's notion was nothing whatsoever. Perhaps this constitutes an example of "history repeats itself." Walcott (1893) produced a significant paper on geologic time that was widely distributed (Yochelson 1989) yet that had elucidated no discussion. The most optimistic interpretation is that the geologic community accepted the interpretation as so satisfactory that no comment was needed.

> The first major bibliography of North American geology indexes Walcott's 1910 "Abrupt Appearance" paper under "Paleontology Cambrian," probably as a consequence of the title. The Lipalian is not separately indexed nor is there any sublisting that would lead to it. In the following bibliography, covering 1919–1928, Lipalian is not indexed. The Lipalian is not listed in the compilation of Wilmarth (1925). It was only just mentioned in a few textbooks and then, except for rare appearances in abstracts, the term has vanished.

In one sense this disappearance is understandable, for though the approach of naming items before there is physical evidence of them has worked well in theoretical physics, probably the idea was never embraced by the geologic community because absence of data did not appeal to geologists. Theoretical concepts for aspects of geology that no longer existed, such as Pangea or Gondwana, were in the literature. They, at least, had some evidence to support their establishment and were grounded in past geography, even if there was no mechanism at the time to explain the observations that supported these notions. Time is ephemeral and basing a concept on lack of data within a continuous sequence is difficult to explain and more difficult to accept.[3]

To reiterate, yes, there is evidence of the "Great Unconformity" in some parts of the world. However, in other parts there are plenty of relevant Precambrian strata, from the period just before the Cambrian. Today that is called the Ediacaran period, and the fossils known from it are *not* thought to represent ancestors to the animals that appear in the Cambrian. In *Darwin's Doubt*, Meyer cites multiple authorities who adopt this view. He also quotes Douglas Erwin and James Valentine who make a crucial point showing why this objection fails:

> In their 2013 book, *The Cambrian Explosion*, paleontologists James Valentine and Douglas Erwin go further. They note that many late Precambrian depositional environments actually provide *more* favorable settings for the preservation of fossils than those from the Cambrian period. As they write, "a revolutionary change in the sedimentary environment—from microbially stabilized sediments during the Ediacaran [late Precambrian] to biologically churned sediments as larger, more active animals appeared—occurred during the early Cambrian. Thus, the quality of fossil preservation in some settings may have actually declined from the Ediacaran to the Cambrian, the opposite of what has sometimes been claimed, yet we find a rich and widespread explosion of [Cambrian] fauna."[4]

The Ediacaran strata, just below the Cambrian, do not yield ancestors to the Cambrian fauna. So the mystery remains.

A paper in *Nature*[5] does try to explain the Cambrian explosion using the Great Unconformity. But it doesn't cite the gap in the fossil record as supposed evidence that Precambrian animal fossils were destroyed. Instead, it makes an even weirder argument—that the weathering of rock, evidence of which we see in the Great Unconformity, dumped a bunch of sediment into the oceans, and *that sediment* triggered the Cambrian explosion. I discussed the paper at *Evolution News & Views*:

> Citing increased chemical weathering around the time of the Cambrian explosion doesn't explain the abrupt appearance of new genes and other genetic information needed to generate new body plans. If they expect us to believe that sedimentation rates explain the sudden origin of new body plans, then it would seem that the Cambrian explosion is still a "mystery."[6]

Compared to this wacky argument, I'd almost prefer the fanciful (and conveniently unfalsifiable) idea that all the Precambrian ancestors disappeared because all their strata weathered away worldwide. That, despite having been discarded by scientists, at least has a veneer of remote plausibility.

Notes

1. Ellis L. Yochelson, "The Lipalian Interval, A Forgotten, Novel Concept in the Geological Column," *Earth Sciences History* 25, no. 2 (2006): 251–269, https://hess. metapress.com/content/772747j430w13n61. See page 265.

2. Ibid., 266.

3. Ibid.

4. Stephen C. Meyer, *Darwin's Doubt: The Explosive Origin of Animal Life and the Case for Intelligent Design* (New York: HarperOne, 2013), 69.

5. Shanan E. Peters and Robert R. Gaines, "Formation of the 'Great Unconformity' as a trigger for the Cambrian explosion," *Nature* 484 (April 19, 2012), 363–66, http://www.nature.com/nature/journal/v484/n7394/full/nature10969.html.

6. Casey Luskin, "Does Lots of Sediment in the Ocean Solve the 'Mystery' of the Cambrian Explosion?," *Evolution News & Views*, April 27, 2012, http://www.evolutionnews.org/2012/04/lots_of_sedimen059021.html.

VI.

TRENDS IN REVIEWING
DARWIN'S DOUBT

*Let me say this as clearly as possible, so
there can be no mistake about what I mean:
there is no controversy. Just because a few
misguided so-called scientists question
the validity of the concept of evolution
doesn't mean there is a controversy.*

GREGORY A. PETSKO, *GENOME BIOLOGY*

Gregory A. Petsko, "It Is Alive," *Genome
Biology*, 9(6) (June 23, 2008): 106.

24.

Evidence of Short-
Term Memory Loss

David Klinghoffer

IF YOU'VE SPENT TIME WITH AGED LOVED ONES AFFLICTED BY SENILE dementia, this will ring a bell. You pay a visit to your old Uncle Ben and in the course of chatting, he comes out with an adamant statement of fact that's clearly in error. Gently, you may try to correct him. But five minutes later, he's forgotten what you said and makes the identical statement, more forcefully than before. You try again but, another five minutes later, you realize the capacity for short-term memory is just no longer there.

In interactions with some of our most adamant critics from the on-line Darwin brigade, it's a lot like that. Well-suited to the defense of an antique of nineteenth-century materialism, there's a certain Darwinian dementia that keeps our interlocutors from assimilating evidence and arguments that go against their views. Some of the responses to Stephen Meyer's book, *Darwin's Doubt*, will illustrate.

It goes like this: They make a claim and you answer them. But shortly after, they are coming at you again with the identical claim, more belligerently than before, as if you'd said nothing at all. Either there's been a genuine memory dump, or they never really heard you, or they did hear and retained what you said, but choose now to act as if they didn't.

Take the matter of the 2003 paper by Long et al. that has served as a rallying cry for Darwinists, supposedly demonstrating, by reviewing many earlier studies, how novel genetic information arises through unguided Darwinian processes.[1] The Long paper, which originally appeared in *Nature Reviews Genetics*, featured prominently in the *Kitzmiller* trial and decision, and our old friend Nick Matzke has been flogging it for years.

You may recall Matzke's review of *Darwin's Doubt*[2] published at the group blog *Panda's Thumb*. This is the one where he supposedly read Meyer's book the day of its publication and immediately composed a 9,400+ word "review" all in little more than 24 hours. (See Casey Luskin's analysis of this feat in Chapter 3.) In Chapter 11 of the (lengthy) book that Matzke purported to read and review in that time, Meyer elaborately, devastatingly debunks the relevance and significance of the Long paper. Meyer cites Matzke by name in Chapter 11, and rebuts his specific arguments regarding Long et al. If you know Matzke as we do, you would expect him to try to bring the hammer down in reply.

Matzke in his response does indeed build up to Meyer's Chapter 11—he must know it's important. But when it comes, it's a feather's blow, amounting to little more than a few bullet points, a spray of snarky insults, *yet another assertion* that Long et al. have it all figured out, and a shrug: "Anyway, most of this has been rebutted elsewhere on [*Panda's Thumb*], and there is little point in doing it again." The tone is so superior and obnoxious that the careless reader might come away with the mistaken idea that Matzke has replied to Meyer.

But he hasn't. It's Uncle Ben, hopping back onto the rumbling freight car of his previous line of thought, undeterred by attempts to set him straight on the facts. "So as I was saying," intones old Ben, paraphrasing Matzke, "all this intelligent design nonsense was refuted way back in 2003 by Dr. Long, and Stephen Meyer never said a word worth repeating in answer to him."

At least Nick Matzke acknowledges that something has been said—though, notwithstanding his high-speed reading of *Darwin's Doubt*, what exactly it was that was said seems mostly to have fallen victim to a memory reset. You can engage geologist Donald Prothero on a subject—in his case, the duration of the Cambrian explosion—only to realize that nothing at all you told him has stuck in his memory. It is truly as if you never opened your mouth.

In *Darwin's Doubt*, Meyer recounts a debate held in 2009, pitting him against Dr. Prothero, where Prothero contended that the Cambrian "explosion" was no explosion at all since it covered a period some 80 million years in length. That's a lot longer than the 5 million years normally cited by Meyer and others.

In the book, Meyer patiently explains, as at the debate, that the figure of 80 million years is arrived at by sleight of hand, a definitional trick:

> As I was listening to [Prothero's] opening statement, I consulted his textbook to see how he had derived his 80-million-year figure. Sure enough, he had included in the Cambrian explosion three separate pulses of new innovation or diversification, including the origin of a group of late Precambrian organisms called the Ediacaran or Vendian fauna. He also included not only the origin of the animal body plans in the lower Cambrian, but also the subsequent minor diversification (variations on those basic architectural themes) that occurred in the upper Cambrian. He included, for example, not just the appearance of the first trilobites, which occurred suddenly in the lower Cambrian, but also the origin of a variety of different trilobite species later from the upper Cambrian.
>
> In my response to Prothero, I noted that he was, of course, free to redefine the term "Cambrian explosion" any way he liked, but that by using the term to describe several separate explosions (of different kinds), he had done nothing to diminish the difficulty of explaining the origin of the first explosive appearance of the Cambrian animals with their unique body plans and complex anatomical features. Beyond this... the Vendian organisms may not have been animals at all, and they bear little resemblance to any of the animals that arise in the Cambrian.[3]

Yet in Prothero's comments on *Darwin's Doubt*, he bluntly reverts to the 80-million-year figure, as if Meyer had said nothing: "[W]e now know that the 'explosion' now takes place over an 80 m.y. time framework. Paleontologists are gradually abandoning the misleading and outdated term 'Cambrian explosion' for a more accurate one, 'Cambrian slow fuse' or 'Cambrian diversification.'"[4] (The latter claim is also extremely deceiving, as you can see in Chapter 5 of *Darwin's Doubt*.)

Prothero also complains that "[Meyer] wastes a full chapter on the empty concept of 'information' as the ID creationists define it,"[5] as if he has forgotten that scientists quite apart from ID advocates have been thinking about the concept of biological information since 1953 when Watson and Crick sparked the realization that storing such information is exactly what DNA does. Meyer explains further in *Signature in the Cell*, but that too has been purged from the mind of Donald Prothero.

Obviously I am not saying these writers suffer from a medical condition, but their brains do appear to have undergone repeated cleansings of what an ordinary scholar, specializing in a relevant field, ought to be able to retain in his mind, especially given frequent reminders.

All of us, to some degree, are selective in the data we choose to assimilate, especially when we're challenged on points of personal significance to us. This is how people maintain their most rigid, fixed ideas against all evidence to the contrary. With Darwinists, what stands out is the combination of arrogance with the failure to register what your opponents have said. As a tactical matter, it sure is convenient for them.

Notes

1. Manyuan Long et al., "Chromosome Rearrangement and Transportable Elements," *Annual Review of Genetics* 36 (2002): 389–410.

2. Nick Matzke, "Meyer's Hopeless Monster, Part II," *Panda's Thumb*, June 19, 2013, http://pandasthumb.org/archives/2013/06/meyers-hopeless-2.html.

3. Stephen C. Meyer, *Darwin's Doubt: The Explosive Origin of Animal Life and the Case for Intelligent Design* (New York: HarperOne, 2013), 72–73.

4. Donald Prothero, "Steven Meyer's Fumbling Bumbling Cambrian Amateur Follies," *Amazon.com*, July 21, 2013, http://www.amazon.com/review/R2HNOHERF138DU.

5. Prothero, "Fumbling Bumbling."

25.

REVIEWING THE REVIEWERS:

A TAXONOMY OF EVASION

David Klinghoffer

FOR HIS REVIEW IN *SCIENCE* OF STEPHEN MEYER'S *DARWIN'S Doubt*, UC Berkeley paleontologist Charles Marshall deserves a prize. Almost alone among critics of the book, Marshall grapples with the main evidence and arguments.

The rest of the hostile reviewers, Darwinian scientists and others, have been... interesting. I've read a lot of book reviews in my professional life. After college I went to work at *National Review* where soliciting, receiving, accepting, rejecting, and editing book reviews, from academics and journalists, was what I did for a decade. It was an education.

One thing I learned is not to be intimated by professors, in science or the humanities. Once you've seen their output in the raw, before being edited, you are never the same again.

Steve Meyer's book tackles a subject that is not only important—what could be more so than the question of whether life's history reflects purpose and intention?—it's also challenging intellectually. While written accessibly for a general audience, it brings together interdisciplinary topics of scholarly study that are not easy.

You look to the men and women who work in the relevant fields for their response. Unfortunately, among critics of *Darwin's Doubt*, the overwhelming unifier has been a consistent evasiveness as to the actual contents of the book. Marshall's review in *Science* is the exception; overall, the reviewers from the field of evolutionary biology and its allied disciplines disappoint me.

Hostile responses of Meyer's book, from Darwinian scholars and the Internet evolution activists who love them too much, fall into the following categories, producing a taxonomy of evasion:

1. The Review Based on Undisguised Ignorance

It started months before publication of *Darwin's Doubt*, when of course no one had read the book, with Jerry Coyne (University of Chicago) and Joe Felsenstein (University of Washington) reassuring blog readers that they already pretty much knew what Meyer's arguments would be. Coyne's classic summary of Meyer's book: "Yes, baby Jesus made the phyla!"[1]

2. The Review Based on Disguised Ignorance

Nick Matzke supposedly read and reviewed *Darwin's Doubt*, in a post of more than 9,400 words published at *Panda's Thumb* little more than 24 hours after he purchased the book on the date of its publication. The suspicion that Matzke failed to give Meyer's book a fair examination, the kind where you read the words on the pages rather than just flipping through and looking for your name in the index, is supported by the fact that he ignored its most important arguments. This didn't stop other writers (Jerry Coyne, Gareth Cook in *The New Yorker*, John Farrell in *National Review*) from citing Matzke as the authoritative source of their dismissals.

Meyer's previous book, *Signature in the Cell*, received much the same treatment, notoriously from Francisco Ayala (UC Irvine) who has given his name to this procedure. To "Ayala" a book is to review and pan it without having read the work.[2]

3. The Reviewer Who Cannot Remember That His
Objections Have Already Been Answered

Donald Prothero (Natural History Museum of Los Angeles County) is the holotype specimen here, though Nick Matzke falls into the category as well. (See my immediately preceding chapter and Casey Luskin's Chapter 22.)

4. The Reviewer Who Cannot Remember That Previous
Reviews of Darwin's Doubt Have Already Been Answered

John Pieret is only a blogger as far as I can tell, not a scholar in any field. He received congratulations for rounding up hostile reviews of *Darwin's Doubt*—without anywhere acknowledging that we have been assiduously demolishing them pretty much as they appear.[3] Again, it's like these guys experience a kind of memory reset when they come across information they don't like, so that displeasing data is sloughed off almost as soon as it is encountered.

5. The Stalled Review

At *Panda's Thumb*, Richard B. Hoppe (apparently not employed as a scientist but identified by *Wikipedia* as the "holder of a PhD in Experimental Psychology from the University of Minnesota") brandishes "the most ambitious effort" at a review of *Darwin's Doubt*, authored by an "anonymous scientist"—Smilodon's Retreat—not to be confused with the Unknown Comic of *Gong Show* fame.[4] This anonymous reviewer said he was "slogging through" the book section by section. Soon he seems to have stalled. In the end, it was evident he was permanently stuck.[5]

6. The Review or Other Response to the Book
Whose Name One Dare Not Speak

In Chapter 28, Casey Luskin points out an interesting strategy employed by a team writing in *Current Biology*, led by Michael S.Y. Lee (University of Adelaide, Australia). In offering a purported solution to what they call "Darwin's dilemma," they "make reference to 'opponents of evolution,' and critique a very Meyer-esque argument, but... refuse to cite Meyer or *Darwin's Doubt* by name."[6]

So too the commentary article in *Science* accompanying Charles Marshall's review, by M. Paul Smith (Oxford University Museum of Natural History) and David A. T. Harper (Durham University), which offers to clarify the true "Causes of the Cambrian Explosion." No mention of Meyer.[7] On the day it came out, Smith and Harper's article was plugged by Carl Zimmer in the *New York Times*—again, with no reference to the real news peg, Meyer's book.[8]

As Luskin notes, "It's now evident that their previous denials notwithstanding, Darwin defenders have been unnerved by *Darwin's Doubt*." They can't ignore us but they can't speak our name either. Why? Probably they think they would be doing us a favor. Writing at *Why Evolution Is True*, Coyne tries to answer Michael Egnor's takedown at *Evolution News & Views* of his views on free will. Coyne excuses himself for doing so: "The [Discovery Institute] has nothing more to do than attack atheists, evolutionary biologists, and tout its Jesus-soaked books; and I don't feel like giving them hits." Ah, baby Jesus and the phyla, again![9]

For Coyne, by his own account, it's about getting "hits" for your blog. For us, it's about an argument on questions of ultimate significance. For us, the point is to establish the truth. A dialogue involving two parties would be helpful in that. Coyne sees things differently. This may explain why, despite wearing the mantle of Darwinian evolution's most vocal American defender, he persistently ducks from a debate.

Have I missed any taxonomic categories? Undoubtedly. Well, the argument over *Darwin's Doubt* is really just getting started. Kudos to Charles Marshall for manning up and wrestling with Meyer's book.

Notes

1. Lawrence A. Moran, "Darwin Doubters Want to Have Their Cake and Eat It Too," *Sandwalk*, April 18, 2013, http://sandwalk.blogspot.com/2013/04/darwin-doubters-want-to-have-their-cake.html.

2. Francisco Ayala, "On Reading the Cell's Signature," *The BioLogos Forum*, January 7, 2010, http://biologos.org/blog/on-reading-the-cells-signature. Compare with: Jay W. Richards, "Ayala: "For the Record, I read *Signature in the Cell*," *Evolution News & Views*,

June 4, 2010, http://www.evolutionnews.org/2010/06/ayala_for_the_record_i_read_si_1035371.html.

3. John Pieret, "Open Wide," *Thoughts in a Haystack*, September 1, 2013, http://dododreams.blogspot.com/2013/09/open-wide.html.

4. "Teach the Controversy," *Wikipedia*, http://en.wikipedia.org/wiki/Teach_the_Controversy; Richard B. Hoppe, "Slaying Meyer's Hopeless Monster," *Panda's Thumb*, September 4, 2013, http://pandasthumb.org/archives/2013/09/slaying-meyers.html.

5. Smilodon's Retreat (pseudonym), "Darwin's Doubt—A Review," *Skeptic Ink*, July 9, 2013, http://www.skepticink.com/smilodonsretreat/2013/07/09/darwins-doubt-a-review/.

6. Casey Luskin, "Teamwork: *New York Times* and *Science* Magazine Seek to Rebut Darwin's Doubt," *Evolution News & Views*, September 24, 2013, http://www.evolutionnews.org/2013/09/teamwork_new_yo077071.html.

7. M. Paul Smith and David A. T. Harper, "Causes of the Cambrian Explosion," *Science* 341, no. 6152 (September 20, 2013): 1355–56, http://www.sciencemag.org/content/341/6152/1355.short.

8. Carl Zimmer, "New Approach to Explaining Evolution's Big Bang," *New York Times*, September 19, 1913, http://www.nytimes.com/2013/09/20/science/new-approach-to-explaining-evolutions-big-bang.html.

9. Jerry Coyne, "Egnorance of free will," *Why Evolution Is True*, October 2, 2013, http://whyevolutionistrue.wordpress.com/2013/10/02/egnorance-of-free-will/. Coyne is responding to Michael Egnor's, "Jerry Coyne Endorses Free Will (Inadvertently as You Might Expect)," *Evolution News & Views*, September 29, 2013, http://www.evolutionnews.org/2013/09/jerry_coyne_end077221.html.

26.

HOSTILE RESPONSES CHANGE

A THOUGHTFUL READER

Casey Luskin

A BUMPER STICKER I'VE SEEN AROUND IN SEATTLE PROTESTS THE War on Terror, warning that "We're making enemies faster than we can kill them." Without wading into matters of national defense and military strategy, I'll give the author of the slogan this much: Any strategy that focuses too much on attacking people, and not enough on making reasoned arguments, is doomed to fail in winning hearts and minds.

For an illustration, take a look at a post by Reverend James Miller, of Glenkirk Church in Glendora, CA. He explains that he became a Darwin skeptic not just after reading *Darwin's Doubt*, but also after considering responses from critics of the book. Under the title "Changing My Mind on Darwin," Pastor Miller writes:

> I've just read Stephen Meyer's *Darwin's Doubt*. Meyer is a Cambridge PhD in philosophy of science. He hangs out with the Intelligent Design people. His writing is fluid, detailed, and reasonable. He seems to know what he's talking about.

> The book makes the case for the fact that the fossil record doesn't support Darwinism. The sudden appearance of new phyla without sufficient time for the mutation and selection process to work is simply unaccounted for by the rocks.

The problem is that when Meyer says things like, "the Precambrian fossil record simply does not document the gradual emergence of the crucial distinguishing characteristics of the Cambrian animals," how on earth should I know if he's right? I don't have time to immerse myself in paleontology. I'll never be an expert. I just have four hundred pages of articulate, self-assured, well-documented evidence for Meyer's case.[1]

Pastor Miller conveys a sentiment that I think is quite reasonable and fairly common. The debate over Darwinism can be technical and complex. Proponents and opponents of neo-Darwinian theory alike cite evidence for their cases. If you haven't had the opportunity to study the scientific questions in detail, it can be difficult to know who is right. If you're not an expert in the science, how can you make an informed decision?

Pastor Miller explains that when he enters a complex debate, he seeks to read arguments from different views. He looks at the evidence, but he also tries to determine who is sincere and credible. Does one side make serious arguments, while the other persistently resorts to personal attacks and name-calling? If so, that can tell you something. Miller explains that he seeks to understand who is behaving as if the evidence is on their side, and who is trying to compensate for a weak position:

> So here's how I find my way into a conversation on subjects that are not my primary field of study. I read the reviews that are antagonistic to the source and just look at the logic that's employed. I find that this often gives me the best read on a work. If the critics are sincere, the reviews are usually precise.

> The *New Yorker*'s review began with a genetic fallacy, presented arguments that Meyer had refuted without mentioning that Meyer had addressed them, and then deferred to another blogger for the scientific content of the review. It then called Meyer "absurd," which, given how shoddy the review actually is, was an absurd thing to do.

> Then I read the review from which the *New Yorker* piece got its "science," which was actually written by a grad student at Berkeley. Now I have to say that Berkeley is, in fact, one of my fields of expertise, and I know exactly how Berkeley grad students go

about their "work." Somehow Berkeley selects the crazies and the militants who show the most promise and then teaches them that knowledge is a completely subjective power tool which should be manipulated by those on an ideological crusade to undermine authority. I'm not kidding. I went to Berkeley. That's what we did.

What's interesting about the grad student's review is that it was posted 24 hours after the release of Meyer's book, and it's filled with snark. He's not having an intelligent conversation, he's insulting Meyer in order to defend something religiously. In a later, defensive review, the grad student says that he read the book "during lunch." He read over 400 pages of scientific material during lunch, and then posted an insulting review. He says his detractors are just "slow readers." People who win speed-reading competitions tend to cover 1,000 words per minute (maybe 4 pages) with 50% comprehension. That level of comprehension is almost useless, and it becomes less useful the more information-rich the content. A book of Meyer's size would have taken an hour and forty minutes at that pace, with minimal retention, and that's if you're not, oh, say, eating lunch. On top of that, the review is almost 10,000 words long, which would take some time to write, making it highly suspicious that the review was written after the book was read and not before, in anticipation of the book's release.

See, this is how I know who to trust in academic communities. The charlatans have no character. You read the grad student's defenses of his review (and they sound a little panicked), and you realize that he has been following Christians around and arguing with them for years with an inquisitor's zeal. There's a personal agenda here, and his approach to new information on the subject is anything but scientific.[2]

That "grad student," of course, is Nick Matzke, who subsequently went to Pastor Miller's blog in an apparent attempt to deconvert Miller from Darwin-skepticism. (To be fair to Nick, besides lunchtime, he claims that he allowed himself "snippets of the afternoon... and then most of the rest of it that night and the next morning"[3] for reading and digesting this massively documented book. Not that that alleviates the problem much.)

You might expect that if your own incivility was the cause of someone's turning away from a viewpoint you want to advance, then you'd try to win them back by being civil and making a respectful, strictly fact-based appeal. If so, then you're not Nick Matzke. That's not how Darwin-defenders think. When confronted with the reality that their style of argument is actually turning people off, Darwin lobbyists often double-down on the nasty rhetoric, evidently thinking the problem was that they *weren't harsh enough to begin with.* Thus Matzke wrote in response to Miller:

> If one is already familiar with the science, it's pretty annoying to see someone like Meyer come in, do a totally hack job which misunderstands or leaves out most of the key data, statistical methods, etc., and then declare that the whole field is bogus. That's why critics are annoyed. And, it's annoying to see other conservative evangelicals blindly follow in his footsteps. Sometimes I think an intelligent design person could say that the idea that the moon is made of rock is a Darwinist conspiracy, and you guys would believe him.[4]

So ID proponents are conspiracy theorists who might say the moon isn't made of rock? And Pastor Miller follows those crazy people? Nick Matzke must think that the best way to bring people over to your side is by demonizing and bullying them—the more, the better. Pastor Miller had a fitting response:

> Actually, Nick, I read Meyer, and you're misrepresenting him through flippant rhetoric rather than simply engaging the facts. You and I both know that he didn't "declare that the whole field is bogus." And your insistence on mischaracterizing his work is a sign that you're not confident that the facts alone discredit him.
>
> As opposed to folly, following the motives and methods of debaters gives you real psychological insight on what they're trying to accomplish, and the scientific enterprise has always prided itself on its objectivity, something we haven't seen from you.

I have the sense that you are actually a brilliant mind. Balance it with character and humility and you'll have far more credibility. I personally would be glad to hear what you have to say if I didn't have to wade through the disrespect.[5]

This recalls the old saying, "When the facts are on your side, pound the facts. When the facts aren't on your side, pound the table." People know this intuitively. Pastor Miller is discerning enough to see how Nick Matzke's disrespect and table-pounding shows that Matzke's viewpoint has a problem with the facts.

Notes

1. James Miller, "Changing My Mind on Darwin," *Hardwired*, October 12: 2013, http://pastorjamesmiller.com/2013/10/12/changing-my-mind-on-darwin/.

2. Ibid.

3. Ibid.

4. In the comments to Miller's "Changing My Mind."

5. Ibid.

VII.

REPLYING TO
DARWIN'S DOUBT
WITHOUT NAMING IT

*There is no significant controversy within
the scientific community about the validity
of the theory of evolution. The current
controversy surrounding the teaching
of evolution is not a scientific one.*

AMERICAN ASSOCIATION FOR THE
ADVANCEMENT OF SCIENCE

American Association for the Advancement of Science Board
of Directors, "Statement on the Teaching of Evolution,"
AAAS.org, February 16, 2006, http://www.aaas.org/sites/
default/files/migrate/uploads/0219boardstatement.pdf.

27.

CAMBRIAN ANIMALS?

JUST ADD OXYGEN

Evolution News & Views

As IF SEEKING TO PREEMPT STEPHEN MEYER'S BOOK *DARWIN'S Doubt,* Darwinian scientists scrambled to publish new explanations of the Cambrian explosion.

Six biologists from Harvard and Scripps (and one from India) lobbed an entry into the contest to explain the Cambrian explosion from Darwinian assumptions. It is "Oxygen, ecology, and the Cambrian radiation of animals," published in *PNAS.* We're not sure if this one passes the laugh test.

They get one thing right: most animal body plans appeared in a geological instant, and no Darwinian evolutionist has ever explained it:

The Proterozoic-Cambrian transition records the **appearance of essentially all animal body plans** (phyla), yet to date **no single hypothesis adequately explains both the timing of the event and the evident increase in diversity and disparity....**

Cambrian fossils chronicle the **appearance of essentially all high-level animal body plans,** as measured by cumulative first appearances of metazoan phyla and classes, in a **geologically brief interval** between ~540 and 500 million years ago.[1]

Oh, they mention *attempts* to explain it, citing external forces such as rising oxygen levels. This can explain the timing (they think), but not the innovation itself:

> There is **no theoretical reason why ocean redox should generate the evolutionary novelties**—specifically the **fundamental new bauplans**—seen in the Cambrian fossil record.[2]

For another, there are ecological hypotheses that focus on interactions among organisms. These can explain the innovation (they think), but not the timing:

> They can also explain the **origin and maintenance of high-level body plan disparity** through the **principle of frustration:** organisms optimally suited to one task will be less well suited for another, leading to a **roughening of the fitness landscape** and isolation of distinct fitness peaks....
>
> A **carnivory-based ecological hypothesis,** then, can **explain the pattern of morphological diversification seen in the Cambrian fossil record** but does not directly address its **timing.**[3]

In other words, the explosion could have occurred tens or hundreds of millions of years earlier. Why did it go bang about 540 million years ago?

What is the secret sauce these researchers have cooked up? Why, it's a mixture of the two competing approaches into an "**integrated causal hypothesis**" that gives credit to both ecological and environmental factors. Here's the scenario: after a billion years of microbe evolution, oxygen levels on the ocean floor rise to "allow" or "permit" higher energy organisms to evolve. Once they evolve, "carnivory" begins. Once carnivory begins, the carnivores need bigger bodies to hunt and eat prey. The evolutionary arms race is on!

> Rising **oxygen** levels would have **allowed** larger body sizes, **but more importantly from a macroevolutionary standpoint,** the first active, **muscular** carnivores... **Escalatory arms races** driven by **these newly evolved carnivores** could then **explain** the relatively **rapid expansion of metazoan diversity and disparity** near the beginning of the Cambrian Period.[4]

Where did these "newly evolved carnivores" (with lots of muscles) come from, you ask? Well, they just appeared:

> Consistent with this hypothesis, the **origin** of carnivory itself **appears** to be temporally correlated with the Proterozoic-Cambrian transition (Fig. 1), a **prerequisite** if predator-prey **"arms races"** are **to be viewed** as the **driving forces** behind **morphological innovation.**[5]

So, carnivores appeared, flexing their muscles. They had to. "Morphological innovation" needed a driving force. Along with muscles, evolution provided new digestive tracts, senses, behaviors—whatever "morphological innovation" the carnivores needed. The hidden hand of evolution had lots of time to prepare for their advent:

> This focus does not obviate a **role** for developmental genetics, but because most gene families that **govern** bilaterian development originated well before Cambrian body **plan** diversification, the prime **role** of development was in **assembling** these preexisting genes into **coherent networks** to **build** body **plans suited** to the evolving Cambrian fitness landscape.[6]

The bold words in the passage above help themselves to the vocabulary of intelligent design. Darwinism does not allow roles. It doesn't permit governors. It doesn't assemble networks. Natural selection is incapable of building things according to a plan. From a Darwinian standpoint, the passage is utterly incoherent.

Every scientific paper needs some kind of appeal to evidence obtained by observation, however obscure it might be. These scientists looked at *modern* sea-bottom communities with low-oxygen conditions and, after sufficient tweaking of definitions, counted carnivores. They only counted polychaetes—annelid worms with feathery appendages for either filtering or catching food—dodging the embarrassing fact that polychaetes are among the novel body plans that appeared in the Cambrian explosion.

Sure enough, they found a trend toward more carnivory with increasing oxygen. There was significant overlap in the data, though, and some significant outliers that bucked the trend. Whether twenty-first century polychaetes have anything to tell about Cambrian ecology, however, is questionable:

> Although these analyses focus on oxygen, **we recognize that other environmental parameters** and physiological stressors **may be important** in **shaping the biology** of modern OMZs [oxygen minimum zones]…

> **Other factors** besides carnivory and oxygen **may** have been important, but many of them are related to **carnivore evolution itself** (e.g., the **evolution of sensory apparatus and vision**).[7]

Hold on. Stop right there. The "evolution of sensory apparatus and vision"? How did that happen?

Once again, we see Darwinists dodging the main problem with the Cambrian explosion: the sudden appearance of biological *information* necessary to build tissues, organs, limbs, eyes, systems, and body plans. This is the focus of most of Part II of *Darwin's Doubt*. Mystically, they imagine animals as eager to evolve but, like racehorses at the gates, held back by environmental barriers:

> Such an **environmental shift** could **remove a barrier to animal evolution,** but aside from direct links to maximum permissible body size, it **lacks an explicit mechanism to generate diversity (new species) and disparity (new body plans).**[8]

That sentence shows that they know better. They know one cannot assume that an opportunity to evolve will generate innovation. One cannot assume that an environmental or ecological "trigger" will propel animals on the race up Mount Improbable. Evolution lacks a *mechanism* to generate body *plans*. The information required to plan and build complex animals is a fundamental challenge to those who trust in unguided, aimless, purposeless natural processes.

Intelligence, by contrast, can assemble codes into coherent networks. Intelligence can direct elements into roles in a hierarchy. We never see unguided processes achieving such ends. From our uniform experience, we can reason that intelligence was required to "govern" new cell types into their "roles" as tissues, organs, and systems in an overall body plan.

A tradition in papers like this is to tag on a promissory note at the end. The researchers present a hint of a suggestion of a notion of a possibility of an idea, about which further research is needed:

> Continued exploration of the causes, timing, and magnitude of oxygenation will provide further insight into the role of oceanographic change in the evolution of carnivory and this unique geobiological event. Further study of the relationship between feeding ecology and oxygen in modern OMZs, as well as the co-evolutionary history of animals and ocean redox state in deep time, may also help us predict future changes associated with ocean deoxygenation and expanding oxygen minimum zones.[9]

"Further insight" presupposes that insight greater than zero has already been offered, but it hasn't. This paper is less about the evolution of carnivory than about the evolution of excuses for maintaining Darwinism despite the evidence of the fossil record—a problem of which Darwin himself was painfully aware, that has only grown worse in the 154 years since *The Origin*.

Notes

1. Erik A. Sperling et al., "Oxygen, ecology, and the Cambrian radiation of animals," *Proceedings of the National Academy of Sciences, USA* 110 (August 13, 2013): 13446–13451, http://www.pnas.org/content/110/33/13446.short. Emphasis added.

2. Ibid. Emphasis added.

3. Ibid. Emphasis added.

4. Ibid., 13449–50.

5. Ibid., 13446. Emphasis added.

6. Ibid., 13450. Emphasis added.

7. Ibid., 13448, 13450. Emphasis added.

8. Ibid., 13446. Emphasis added.

9. Ibid., 13450.

28.

TEAMWORK: *NEW YORK TIMES* AND SCIENCE OFFER A REBUTTAL

Casey Luskin

IT'S NOW EVIDENT THAT, THEIR PREVIOUS DENIALS NOTWITHSTAND-
ing, Darwin defenders have been unnerved by *Darwin's Doubt*. On the
same day, September 20, 2013, both the world's top newspaper (the *New
York Times*) and one of the world's top scientific journals (*Science*) turned
their attention to the problem posed by Stephen Meyer. The review of
Darwin's Doubt in *Science* was by Charles Marshall. Let's take a look at
science-writer Carl Zimmer's piece in the *Times*, "New Approach to
Explaining Evolution's Big Bang."[1] Zimmer promotes the conclusions of
a commentary article—published in the same issue of *Science* as Marshall's review of Meyer's book—which purports to explain the Cambrian
explosion.[2]

There's something odd about Zimmer's article. Despite the vigorous media dialogue over *Darwin's Doubt*, in print and online, Zimmer
declines to mention the book or its author. But then the article in *Science* that claims to reveal the causes of the Cambrian explosion never
acknowledges the controversy either. *ENV* noted a similar reticence[3] in
a *Current Biology* paper, which makes reference to "opponents of evolu-

tion,"[4] and critiques a very Meyer-esque argument, but likewise refuses to cite Meyer or *Darwin's Doubt* by name.

Zimmer endorses an approach to the Cambrian explosion, taken by M. Paul Smith and David A. T. Harper who wrote the *Science* commentary, that's often seen in papers on the subject. These papers cite a myriad of explanations, on the apparent assumption that just by tossing out a bunch of scattershot ideas, you've solved the problem. Carl Zimmer describes the method as follows:

> Geologists suggested geological causes. Ecologists proposed ecological ones. Many of those ideas have merit, Dr. Smith and Dr. Harper argue in a commentary in this week's *Science*, but it's a mistake to search for a single cause. They propose that a tangled web of factors and feedbacks were responsible for evolution's big bang.[5]

How did that work? Zimmer writes:

> Long before the Cambrian explosion, Dr. Smith and Dr. Harper argue, one lineage of animals had already evolved the genetic capacity for spectacular diversity. Known as the bilaterians, they probably looked at first like little crawling worms. They shared the Precambrian oceans with other animals, like sponges and jellyfish. During the Cambrian explosion, relatively modest changes to their genes gave rise to a spectacular range of bodies.

> But those genes evolved in bilaterians tens of millions of years before the Cambrian explosion put them to the test, notes Dr. Smith. "They had the capacity," he said, "but it hadn't been expressed yet."[6]

Isn't that interesting—bilaterians "evolved the genetic capacity for spectacular diversity," for no apparent reason, long before it was "expressed." The *Science* paper notes "an apparent >100-million-year gap between the evolutionary innovation and its consequences"! For all that time, the "genetic capacity" sat on its hands, doing nothing. Then, thanks to sheer dumb luck, it turned out that the "innovation" was exactly what was necessary to evolve into all the diverse forms of animals we observe. The only thing missing was an environmental trigger.

The trouble is that, in Darwinian theory, you don't survive and reproduce based upon what will happen in the future. You survive and reproduce based upon what happens now. Darwinian evolution can't select for future goals, and thus could not evolve the "genetic capacity for spectacular diversity" in the future. Despite their theory, which was formulated to explain away the appearance of teleology in biology, Darwinians are being forced into increasingly teleological-sounding explanations for the Cambrian explosion. Not that Team Darwin is anywhere near to admitting that.

As Meyer explains in *Darwin's Doubt*, building new forms of animal life requires massive amounts of new biological information in the form of myriads of new genes, non-coding DNA regulatory elements, gene regulatory networks, and epigenetic information. He shows, for several separate reasons, that the neo-Darwinian mechanism lacks the creative capacity necessary to generate these various forms of information.

For example, Meyer shows that functional genes and proteins are exceedingly rare within sequence space. And, for this reason, he argues that a random mutational search will be overwhelmingly more likely to fail, than to succeed, in generating even a single new gene or protein during the entire history of life on Earth. Similarly, he shows that mutations in DNA alone cannot produce the epigenetic ("beyond the gene") information necessary to build new animal body plans.[7]

Does Zimmer, or the article in *Science* that he cites, address (or solve) these or any of the other problems that Meyer addresses? No.

ID theorists pay close attention to the crucial question: *Where does the information necessary to build a new animal come from?* Zimmer and the scientists he writes about don't even ask that question.

They just assume the "genetic capacity" arose 100+ million years before it was "expressed"—without providing any causal explanation for the origin of that information. In other words, they just assume an animal with all the necessary information to produce all future Cambrian animals. That's quite an assumption! Of course, once that information had arisen, all that was then required was some global environmental change to trigger "an evolutionary cascade that led to the rapid rise in diversity" (as the *Science* paper puts it). Because Earth's history is filled with geological changes and environmental catastrophes, such events aren't hard to find. Indeed, they're practically a dime a dozen. Here's what Zimmer finds:

> It took a global flood to tap that capacity, Dr. Smith and Dr. Harper propose. They base their proposal on a study published last year by Shanan Peters of the University of Wisconsin and Robert Gaines of Pomona College. They offered evidence that the Cambrian Explosion was preceded by a rise in sea level that submerged vast swaths of land, eroding the drowned rocks.[8]

I have responded to Peters and Gaines's study at *Evolution News & Views*, twice. Because, puzzlingly, it continues to be cited, over and over. As I wrote:

> Citing increased chemical weathering around the time of the Cambrian explosion doesn't explain the abrupt appearance of new genes and other genetic information needed to generate new body plans. If they expect us to believe that sedimentation rates explain the sudden origin of new body plans, then it would seem that the Cambrian explosion is still a "mystery."[9]

Wait, there's more. "But these great floods also poisoned the ocean," Zimmer says, and "In order to survive, animals had to evolve ways to rid themselves of the poison." Are we about to hear an explanation for how new information arose? No:

> One solution may have been to pack the calcium into crystals, which eventually evolved into shells, bones, and other hard tissues. Dr. Smith doesn't think it's a coincidence that several different lineages of bilaterians evolved hard tissues during the Cambrian explosion, and not sooner.[10]

According to this logic, increasing the level of "poison" (calcium) in water generates new information. From there, it's a snap:

> These shells and other hard tissues sped up animal evolution even more. Predators could grow hard claws and jaws for killing prey, and their prey could evolve hard shells and spines to

defend themselves. Animals became locked in an evolutionary arms race.[11]

OK, I think I now understand why the Cambrian explosion happened. Here's the formula:

- First, the "genetic capacity" to produce all known animal forms arises without any adaptive benefit in some unknown hypothetical ancestral organism.

- Then it does nothing for some 100+ million years. (Nobody's sure exactly how long.)

- Then some environmental trigger adds selection pressure. Earth's history is full of options; choose one, or choose five. Zimmer's scientists choose chemical weathering + sea level rise + oxygenation of oceans.

- Then an "arms race" ensues, and all that untapped genetic information is suddenly "expressed," and boom goes the dynamite: numerous animal body plans appear in a geological blink of the eye.

The *Science* commentary puts it more artfully: "Together, these interacting processes generated an evolutionary cascade that led to the rapid rise in diversity."[12] And so, there you have it: Cambrian enigma solved—provided, of course, that you don't ask any pesky questions about the origin of genetic or epigenetic information.

Notes

1. Carl Zimmer, "New Approach to Explaining Evolution's Big Bang," *New York Times*, September 19, 2013, http://www.nytimes.com/2013/09/20/science/new-approach-to-explaining-evolutions-big-bang.html.

2. M. Paul Smith and David A. T. Harper, "Causes of the Cambrian Explosion," *Science* 341, no. 6152, (September 20, 2013), http://www.sciencemag.org/content/341/6152/1355.

3. "How to Solve the Cambrian Explosion: Turn Up the Evolutionary Speed Dial," *Evolutionary News and Views*, September 20, 2013, http://www.evolutionnews.org/2013/09/how_to_solve_th076861.html.

4. Michael S. Y. Lee, Julien Soubrier, Gregory D. Edgecombe, "Rates of Phenotypic and Genomic Evolution during the Cambrian Explosion," *Current Biology*, 23

(October 7, 2013): 1889–1895, http://www.cell.com/current-biology/abstract/S0960-9822(13)00916-0.

5. Zimmer, "New Approach."

6. Ibid.

7. Stephen C. Meyer, *Darwin's Doubt: The Explosive Origin of Animal Life and the Case for Intelligent Design* (New York: HarperOne, 2013), 271–87.

8. Zimmer, "New Approach."

9. Casey Luskin, "Does Lots of Sediment in the Ocean Solve the 'Mystery' of the Cambrian Explosion?," *Evolution News & Views*, April 27, 2012, http://www.evolutionnews.org/2012/04/lots_of_sedimen059021.html.

10. Zimmer, "New Approach."

11. Ibid.

12. Smith and Harper, "Causes of the Cambrian Explosion."

29.

To Create Cambrian Animals, Whack the Earth from Space

Evolution News & Views

IT'S SURELY NOT A COINCIDENCE THAT THE SEASON IN SCIENCE-journal publishing in which *Darwin's Doubt* was released saw a variety of attempts to solve the enigma that Stephen Meyer describes in the book. The problem, of course, is how to account for the geologically sudden eruption of complex new life forms in the Cambrian explosion. Meyer argues that the best explanation is intelligent design.

The orthodox materialist camp in mainstream science remains in full denial mode. They can't stomach the proposal of ID, but neither can they for the most part bring themselves to answer Meyer by name, or even admit there's a controversy on the subject. Charles Marshall, reviewing the book in *Science*, is the honorable exception. So we get what look like stealth responses to Meyer's book that claim to have figured out the Cambrian puzzle without telling you what the urgency for doing so really is, thus evading the task of responding to Meyer directly. (See Chapter 25 in this book for David Klinghoffer's "Reviewing the Reviewers: A Taxonomy of Evasion.")

Probably the most hopeless solution so far ascribes some of the creative power to a blast in the ocean by a space impact. This supposedly helped "set the stage" for the rapid proliferation of new animal forms. When we examine the complexity of a single Cambrian fossil, though, such a notion, like the others on offer, leaves all the important questions unanswered.

To his credit, Grant M. Young, the author of the proposal, was somewhat modest in the way he formulated his idea. His paper in *GSA Today* is primarily concerned with looking for evidence of a "very large marine impact" prior to the Ediacaran period that sent vast quantities of water and oxygen into the atmosphere, changed the obliquity of Earth's spin access, and altered sea levels. The aftermath of that catastrophe, he speculates, *played a role* in the Cambrian explosion—but a "crucial" one.

> Attendant unprecedented **environmental** reorganization **may have played a crucial role in the emergence of complex life forms.**[1]

That's all Young had to say about it, but the suggestion was enough for NASA's *Astrobiology* Magazine to jump on it with a breathless headline: "Did a Huge Impact Lead to the Cambrian Explosion?" Author Johnny Bontemps catapulted that tease into the notion that "[t]he ensuing **environmental re-organization** would have then **set the stage for the emergence of complex life.**" Bontemps is correct about one thing:

> These events marked the beginning of **another drastic event known as the Cambrian explosion.** Animal life on Earth **suddenly blossomed,** with **all of the major groups of animals alive today making their first appearance.**[2]

Let's take a look at just one of the Cambrian animals, as seen in an exquisitely preserved new fossil from the Chengjiang strata in China, where so many beautiful fossils have been found. The new fossil, *Alalcomenaeus*, published by *Nature*, was furnished with multiple claws like other Cambrian arthropods, but was so well preserved that its nervous system could be outlined in detail.[3] Even though it is dated from the early Cambrian at 520 million years old, it already had the nerves of modern spiders. Co-author Nick Strausfeld explains:

> "We now know that the megacheirans had **central nervous systems very similar to today's horseshoe crabs and scorpions,**"

said Strausfeld, the senior author of the study and a Regents' Professor in the UA's Department of Neuroscience. "This means the **ancestors of spiders and their kin lived side by side with the ancestors of crustaceans in the Lower Cambrian.**"[4]

Though *Alalcomenaeus* was tiny (about an inch long), its nervous system must have been fairly advanced, because the elongated creature was capable of swimming or crawling or both. In addition to about a dozen body segments with jointed appendages, it had a "pair of **long, scissor-like appendages attached to the head,** most likely for **grasping** or **sensory purposes.**"[5] It also had two pairs of eyes.

Iron deposits had selectively accumulated in the nerve cells, allowing the research team to reconstruct the highly organized brain and nervous system. After processing with CT scans and iron scans, "**out popped this beautiful nervous system in startling detail.**"

Comparing the outline of the fossil nervous system to nervous systems of **horseshoe crabs and scorpions** left **no doubt** that 520-million-year-old *Alalcomenaeus* was a member of the chelicerates.

Specifically, the fossil shows the **typical hallmarks of the brains found in scorpions and spiders:** Three clusters of nerve cells known as ganglia fused together as a **brain** also fused with some of the animal's body ganglia. This **differs from crustaceans** where ganglia are further apart and connected by long nerves, like the rungs of a rope ladder.

Other diagnostic features include the forward position of the **gut opening** in the brain and the arrangement of **optic centers** outside and inside the brain supplied by **two pairs of eyes, just like in horseshoe crabs.**[6]

(Horseshoe crabs survive as "living fossils" to this day, as residents near the Great Lakes know from the annual swarms.)

Alalcomenaeus resembles modern chelicerates, one of the largest subphyla of arthropods, including horseshoe crabs, scorpions, spiders, mites, harvestmen, and ticks. *Live Science* adds, "The discovery of **a fossilized brain** in the preserved remains of an extinct 'mega-clawed' creature has revealed **an ancient nervous system that is remarkably similar to that of modern-day spiders and scorpions.**[7]

Since crustaceans and chelicerates have both been found in the early Cambrian, Darwinian evolutionists are forced to postulate an unknown ancestor further back in time: **"They had to come from somewhere,"** Strausfeld remarks. **"Now the search is on."**[8] That sounds like the same challenge Charles Darwin gave fossil hunters 154 years ago to find the ancestors of the Cambrian animals.

The difficulty? It requires many different tissue types and interconnected systems to operate a complex animal like *Alalcomenaeus*, with its body segments, eyes, claws, mouth parts, gut, and nervous system with a brain, to say nothing of coordinating the developmental programs that build these systems from a single cell. That is the major problem that Stephen Meyer emphasizes in *Darwin's Doubt*: Where does the information come from to build complex body plans with hierarchical levels of organization?

Slamming a space rock at the Earth is hardly a plausible source of information. Stephen Meyer and Casey Luskin have answered in detail the most serious and scholarly critique of his book, by Charles Marshall, refuting Marshall's criticisms point by point.[9] Meanwhile, the proposed alternative explanations for the Cambrian event keep coming, bearing increasingly the marks of desperation.

Notes

1. Grant M. Young, "Evolution of Earth's climatic system: Evidence from ice ages, isotopes, and impacts," *GSA Today* 23, no. 10 (October 2013): 4–10, http://www.geosociety.org/gsatoday/archive/23/10/article/i1052-5173-23-10-4.htm. Emphasis added.

2. Johnny Bontemps, "Did a Huge Impact Lead to the Cambrian Explosion?" *Astrobiology Magazine*, October 14, 2013, http://www.astrobio.net/exclusive/5742/did-a-huge-impact-lead-to-the-cambrian-explosion. Emphasis added.

3. Gengo Tanaka et al., "Chelicerate neural ground pattern in a Cambrian great appendage arthropod," *Nature* 502 (October 17, 2013): 364–67, http://www.nature.com/nature/journal/v502/n7471/full/nature12520.html.

4. Daniel Stolte, "Extinct 'Mega Claw' Creature Had Spider-Like Brain," *UA News*, October 16, 2013, http://uanews.org/story/extinct-mega-claw-creature-had-spider-like-brain. Emphasis added.

5. Ibid. Emphasis added.

6. Ibid. Emphasis added.

7. Denise Chow, "Ancient 'Mega-Clawed' Creature Had Brain Like a Spider's," *Live Science*, October 16, 2014, http://www.livescience.com/40474-ancient-mega-clawed-creature-fossilized-brain.html. Emphasis added.

8. Ibid. Emphasis added.

9. See Chapters 10-14 of *Debating Darwin's Doubt*.

30.

DOES LIGHTNING-FAST EVOLUTION

SOLVE THE CAMBRIAN ENIGMA?

Stephen C. Meyer

IN SEPTEMBER 2013, THE SCIENCE NEWS MEDIA WERE ABUZZ[1] ABOUT a paper in *Current Biology*, "Rates of Phenotypic and Genomic Evolution During the Cambrian Explosion," by Michael Lee and his colleagues. The paper purports to show that rates of evolutionary change during the Cambrian period were elevated—not, however, to such an extent as to upset the neo-Darwinian understanding of evolution via natural selection and random mutation.[2]

The insistence that these findings pose no problem for Darwin's theory has been a major theme in media commentary about the paper. Reporting on its conclusions, *Live Science* explains that "scientists have figured out just how quickly evolution was occurring during evolution's 'big bang.' And it was fast by most measures, five times quicker than occurs today."[3] Dr. Lee is quoted as saying that he finds this "perfectly consistent with Darwin's theory of evolution." A piece at *Science Now*, the journal *Science*'s news desk, goes further, assuring readers that these results not only are reconcilable with but positively vindicate evolution by natural selection: "Their finding—that the rate of change was high, but still plausible—may put Darwin's fears to rest."[4]

This curious reference to the "fears" of the long-deceased Darwin may reflect an implicit acknowledgment of the challenge posed by *Darwin's Doubt*. If so, it would not be the first time that the science media or a science journal has responded to the arguments in the book without referring to it by name. This has become somewhat of a pattern. As Casey Luskin noted, the paper by Lee et al. itself "makes reference to 'opponents of evolution,' and critiques a very Meyer-esque argument."[5]

So then, let's take a closer look. Does this paper in *Current Biology* explain the explosive origin of animal life in the Cambrian period? In other words, does it identify a causal mechanism capable of producing the novel animal forms and biological information that arose during the Cambrian? Does it thus provide a refutation of the main arguments of *Darwin's Doubt*?

It does not. Instead, by using the term "evolution" in equivocal ways, the authors end up presenting the problem of the Cambrian explosion (the rapid emergence of new forms of animal life) as its own solution (which they simply re-describe as the rapid "evolution" of new forms of animal life).

To understand this exercise in rhetorical legerdemain, we need to recall that "evolution" can be defined in several different ways. The term may refer to: (1) the fact of biological change over time, (2) the theory of universal common descent (which implies continuous biological change over time), or (3) the claim that natural selection acting on random variations and mutations is sufficient to *cause* the change that has occurred in the history of life, including major morphological innovation such as occurred during the Cambrian explosion.

As I observe in *Darwin's Doubt*, given a Darwinian commitment to universal common descent—"evolution" in the second sense—the absence of discernible ancestors in the Precambrian fossil record is mysterious. However, my main argument in the book concerns the inadequacy of "evolution" in the third sense. I argue (for five separate reasons) that the mutation/natural selection mechanism lacks the creative power to produce the origin of the new forms of animal life in the Cambrian period.

Does this new paper answer, or even address, the challenges to the creative power of the mutation/selection mechanism? Does it show that the mutation/selection mechanism, or any other undirected materialistic mechanism, could generate the new genetic (and epigenetic) information necessary to produce the innovations in form and structure that occurred in the Cambrian period? It does not. At most it measures the "rate of change" that occurred within one phylum during (and after) its origin in the Cambrian.

The words "rate of change" are key here. Even ignoring the paper's other problems, Lee and his colleagues only succeed in measuring a *rate* at which molecular and morphological change occurred ("evolution" in the first sense) during and after the Cambrian period. The study never established that the change it measured was *caused* by natural selection and random mutations, or any other purely materialistic evolutionary mechanism. Thus, it does not provide a causal explanation for the origin of the animal forms that arise in Cambrian period—the absence of which constitutes the central mystery addressed in *Darwin's Doubt*.

The study begins by using molecular and morphological data to construct phylogenetic trees of arthropods. On the assumption that degree of biological similarity reflects the degree of relatedness, these trees were constructed by comparing the morphological traits and molecular sequences of various *living* arthropod species, and then grouping these species according to their number of shared similarities.

As is common in such studies, the length of a branch on a phylogenetic tree corresponds to the amount of change that presumably took place along that branch. In a tree derived from the comparative analysis of similar DNA sequences in different organisms, branch length corresponds to the number of nucleotide differences in the two respective molecules, and, thus, presumably, to the number of bases that have changed since the two organisms possessing these molecules diverged from a common ancestor. On a morphology-based tree, branch length would correspond to how many morphological characters have changed since the presumed divergence.

The hypothetical phylogenetic tree below illustrates these conceptual relationships, with different branch lengths leading to three fictional living organisms A, B, and C:

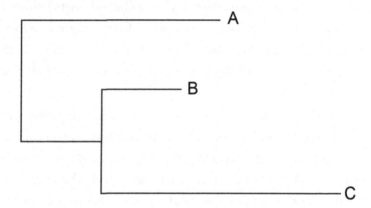

Figure 30-1. Illustration: Jens Jorgenson and Casey Luskin.

In this hypothetical tree, the "branch" length reflects the amount of change that has taken place during the evolution of that organism from its presumed ancestor. Here's the same tree with fictional units of "change" added:

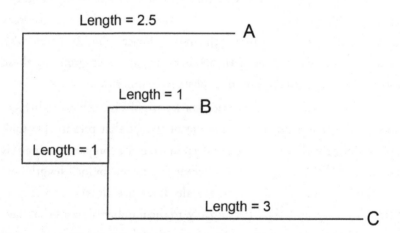

Figure 30-2. Illustration: Jens Jorgenson and Casey Luskin.

Lee and his colleagues applied this method of analysis to arthropods. Compared to many other invertebrates, arthropods have a rich fossil record. So by using fossils to date the nodes (i.e., the starting points and endpoints of branches) on their hypothetical tree, they approximated how long a given branch lasted in real time.

To illustrate further, let's assume that the first representative of a group—one that includes A, B, and C—appears in the fossil record around 400 million years ago. We've now dated the base of their group within the tree, labeled "400 mya" below:

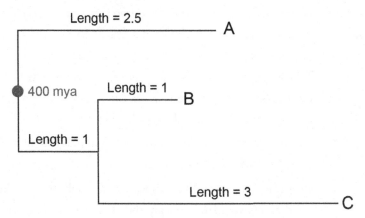

Figure 30-3. Illustration: Jens Jorgenson and Casey Luskin.

Now let's say that the first member of the group that includes just B and C appears in the fossil record at 200 million years ago. Now we can date the split of that group as well:

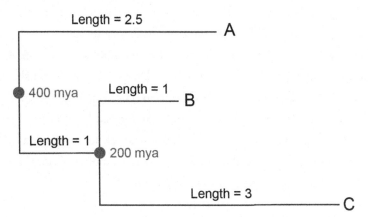

Figure 30-4. Illustration: Jens Jorgenson and Casey Luskin.

Keep in mind that the length of the branch doesn't necessarily correspond to the amount of time elapsed. Rather, it corresponds to the amount of biological change that occurred along that branch (i.e., the number of measured molecular or morphological differences between the two groups representing the beginning and end of the branch). The length of time represented by the branch is only assigned after the fact, using fossils thought to mark the first appearance of the two groups.

Nevertheless, once evolutionary biologists think they know how much change has taken place (the number of molecular or morphological differences) along a branch (between two different organisms), and how much time elapsed along the branch (i.e., between the presumed first appearance of the two organisms in question), they can then calculate a rate of evolutionary change. So now that we've included a couple of fossil dates in our hypothetical tree, we can start calculating rates of change along the branches that led to A, B, and C:

+ Species "A" changed at a rate of 2.5 units/400 mya = 0.6 units/100 million years.

+ After the group including species B and C split from A, but before B and C split from one another, they changed at a rate of 1 unit/200 mya = 0.5 unit/100 million years.

+ But after the split of B and C, B continued to change at a rate of 0.5 units/100 million years, whereas the rate of change in C accelerated to 3 units/200 mya = 1.5/100 million years.

All this is very well. However, methods for calculating rates of change do not establish the *cause* of the change in question. This is axiomatic. And it is just as true of genetic or morphological change as of any other kind. For this reason, the Lee et al. paper did not establish that the emergence of animals in the Cambrian "could be explained... by way of natural selection."[6] Instead, it only established *how much* genetic and morphological change natural selection and random mutation (or some other cause) would need to produce in a given amount of time in order to explain the origin and evolution of arthropods.

After dating the nodes of their hypothetical trees, the Lee et al. study calculated that rates of genetic and morphological change among arthropods during the Cambrian period were five times greater than they were after the Cambrian period. Fair enough. But did the authors establish that mutation and natural selection could generate the amount of change that their study measured? Did they establish that natural selection *was responsible* for the genetic and morphological change that had occurred within arthropods?

They did not. The authors *assumed* that natural selection and random mutations were responsible for the change that had occurred and then simply asserted that natural selection *could* produce the rate of morphological change they measured. In other words, they begged the question as to the rapidity with which the mechanism of mutation and selection can produce morphological novelty. They did not demonstrate that the neo-Darwinian mechanism has the creative power to generate morphological novelty this quickly.

Thus, although Lee and colleagues claim to have refuted unnamed "opponents of evolution," they certainly did not refute the specific quantitative challenges to the creative power of the mutation selection mechanism presented in *Darwin's Doubt*, which cast doubt on the ability of the neo-Darwinian mechanism to produce even modest changes or innovations in single proteins within known evolutionary deep time. Lee did not explain how random mutation and natural selection could have overcome the problem of the rarity of genes and proteins in combinatorial sequence space. Nor did his team show that the waiting times associated with the production of even a few coordinated mutations were any shorter than the prohibitively long waiting times calculated by the researchers cited in *Darwin's Doubt*.

For this reason, the study does not justify the claim of its punch line: "Inexplicably fast rates are not required to explain the Cambrian explosion of arthropods, even under an extreme scenario in which all divergences are compressed into the Cambrian."[7]

The study by Lee et al. is, however, problematic for other reasons.[8]

Notes

1. "How to Solve the Cambrian Explosion: Turn Up the Evolutionary Speed Dial," *Evolution News & Views*, September 20, 2013, http://www.evolutionnews. org/2013/09/how_to_solve_th076861.html.

2. Michael S. Y. Lee, Julien Soubrier, and Gregory D. Edgecombe, "Rates of Phenotypic and Genomic Evolution during the Cambrian Explosion," *Current Biology* 23 (October 7, 2013): 1889–1895, http://www.cell.com/current-biology/abstract/S0960-9822(13)00916-0.

3. Tia Ghose, "Lightning-Fast Evolution Clocked During Cambrian Explosion," *Live Science*, September 12, 2013, http://www.livescience.com/39596-cambrian-explosion-evolution-measured.html.

4. Kelly Servick, "Evolution's Clock Ticked Faster at the Dawn of Modern Animals," *Science*, September 12, 2013, http://news.sciencemag.org/evolution/2013/09/evolution%E2%80%99s-clock-ticked-faster-dawn-modern-animals.

5. Casey Luskin, "Teamwork: *New York Times* and *Science Magazine* Seek to Rebut *Darwin's Doubt*," *Evolution News & Views*, September 24, 2013, http://www. evolutionnews.org/2013/09/teamwork_new_yo077071.html.

6. Lee et al., "Rates of Phenotypic and Genomic Evolution."

7. Ibid.

8. Casey Luskin, "*Current Biology* Paper's Assumptions and Methodology Dramatically Underestimate 'Rates of Change' in the Cambrian Explosion," *Evolution News & Views*, October 31, 2013, http://www.evolutionnews.org/2013/10/current_biology078581. html.

VIII.
RESPONSES FROM THEISTIC EVOLUTIONISTS

[E]volution shares equal status with such established concepts as the roundness of the earth, its revolution around the sun, and the molecular composition of matter.

THEOLOGIAN MICHAEL PETERSON, ASBURY SEMINARY

Michael Peterson, "C. S. Lewis on Evolution and Intelligent Design," *Perspectives on Science and the Christian Faith* (December 2010), 266.

31.

FINDING THE DESIGNER'S GARBAGE

Casey Luskin

I KEEP GOING BACK AND FORTH IN MY MIND ABOUT CAMBRIDGE UNIversity paleontologist and theistic evolutionist Robert Asher. Is he a serious, civil critic of intelligent design, or just another typical Internet Darwin defender who uses and endorses nasty rhetoric and weak objections, often misrepresenting the arguments for ID?

My first encounter with Dr. Asher came in September 2012, when he wrote an article at the *Huffington Post* attacking the textbook *Explore Evolution*. This wasn't an encouraging introduction to Dr. Asher. In the article, he claimed: "This book makes a case that biodiversity results from a kind of 'design' incompatible with evolution by natural selection,"[1] as if the textbook argued for intelligent design.

In reality, *Explore Evolution* doesn't argue for intelligent design, whether explicitly or implicitly. The word "design" is used a few times, but it's in the same context many anti-ID biologists use it: to refer to the structural "design" of an organism, not to argue for intelligent design. At the time I posted my rebuttal to Asher, I challenged him (or anyone else) to provide page numbers and quotes showing just where and how the book argues for intelligent design.[2] He, of course, never took me up on my challenge. His critique of *Explore Evolution* included other egregious misrepresentations of the textbook's arguments, ridiculing us as "anti-science."

I was more encouraged in early 2013, after I read Asher's book *Evolution and Belief: Confessions of a Religious Paleontologist*. Though I disagreed with many of his criticisms of intelligent design, I found that he generally tried to address ID's arguments squarely. I was especially impressed that he acknowledged that Stephen Meyer "claims to use the uniformitarianism of Charles Darwin to justify his inference."[3]

Admittedly in a crude way, his book even accurately stated that Meyer's argument was a positive one that followed from the complexity of living organisms. Though Meyer wouldn't put it quite that way, Asher characterized Meyer's argument as saying that "a very complex device we observe now, such as a wristwatch, computer, or piece of software, has only one source: human ingenuity" and thus "a similar complex device we observe in the geological past must also have arisen as a result of something like human ingenuity, i.e., intelligence."[4] Though Asher's version of the argument wasn't nearly as sophisticated as Meyer's, at least it appeared that Asher had read and understood some of Meyer's writings. True, Asher's book *Evolution and Belief* was marred by citation bluffs[5] and common misunderstandings about ID,[6] but its serious tone was a big improvement on the critiques of many other ID opponents.

Fast forward to 2014. Robert Asher has now critiqued *Darwin's Doubt* in the *Huffington Post*.[7] Sad to say, some progress has been lost with Dr. Asher.

Titled "A New Objection to Intelligent Design," Asher's article opens with the announcement, "I'm not going to review his book, which has already received well-deserved, accurately disparaging coverage by practicing scientists (like Nick Matzke, Don Prothero, and Charles Marshall)." In fact, *Darwin's Doubt* received lots of typical empty ridicule from Matzke and Prothero (Marshall was civil), which I won't dignify by quoting. Matzke and Prothero's reviews were full of gutter rhetoric— "disparaging" treatment indeed, which Asher apparently feels was "well-deserved."

It turns out that Asher's own complaint against *Darwin's Doubt* has very little to do with the science, and is instead a weak quasi-philosophical objection. In a nutshell, Asher argues that since all the intelligent agents in our experience are human agents, and since humans have teeth and bones, and do things like leaving behind waste and garbage, therefore we can't claim to detect design unless we find evidence of the designer's teeth or bones, or waste or other abandoned material, in the historical record.

This is all in the context of a challenge to Stephen Meyer's uniformitarianism. Asher writes: "Meyer is not really uniformitarian in a scientific sense" since "while Meyer appeals to the uniformitarianism of Darwin to address this possibility [that a human-like, super-intelligence interfered with life], he stops short of fully using it." He continues:

> Here is why: if an intelligent force actually seeded the Earth with biological novelties over time (like bipedal apes), uniformitarianism would lead us to expect that intelligence to have left behind a record, in the same way that any other intelligence would leave behind a record. For starters, we'd expect to find hard organic remains such as bones or teeth, since all known intelligent agents have them. Furthermore, if these agents could engineer a new organism, we should reasonably expect them to leave behind some of the more banal traces of their existence, like infrastructure and waste, beyond simply their finished product, such as a new ape. Remains of things derived from human "intelligence" (metal alloys, synthetic polymers, cigarette butts, etc.) will be at least as obvious to future geologists as the global traces of an asteroid impact 65 million years ago are to geologists today...

If we really apply uniformitarianism to determine if intelligent agents influenced the course of our evolutionary history, we'd expect those agents to have left behind the same kinds of traces as other such agents. Humanity is the best example we've got so far, and we make an exponentially greater amount of garbage than we do functional designs. One of the most obvious kinds of material evidence that a human-like intelligence in Earth's distant past would have left behind was spelled out with one of the most famous lines, indeed one of the most famous words, ever uttered in twentieth-century film: Plastics. Far from being persecuted

for a discovery that raises the issue of design, anyone finding genuine "plastic spikes" in deep time, corresponding temporally to
one or more evolutionary events, would be assured of a successful, mainstream academic career (to say the least). While such
artifacts wouldn't tell us how biodiversity actually came about,
they would indicate that something out there served as an agent
behind life on Earth. Maybe ID advocates will claim that their
"intelligence" didn't have to leave behind a plastic spike or other
such material evidence. And when they do, they cease to qualify
as scientifically uniformitarian.[8]

Asher makes much the same argument in his book *Evolution and
Belief*. While he basically concedes that design detection is a theoretical
possibility, his critique is flawed on at least two levels.

First, he's wrong to claim that we must find evidence of the designer's
"waste" or "infrastructure" (or body parts) to detect design.[9] The defining characteristic of intelligence is *not* whether the designer has teeth or
bones, or leaves behind waste or garbage. Rather, the defining feature
is the ability to rationally choose between many options, and look forward with will, forethought, and intentionality to solve some complex
problem. Accordingly, when intelligent agents act, they generate high
levels of complex and specified information (CSI). Thus, a fundamental
sign that an intelligent agent has been at work is high CSI. We can use
perfectly legitimate uniformitarian reasoning to detect design by finding
high CSI, regardless of whether we also find physical evidence of the
designer's body, waste, or infrastructure.

Second, Asher is wrong to claim that we haven't found any evidence
that the intelligent designer has left a record. In fact we find all kinds of
"counterflow" in biology—in the form of polymers (proteins) and computers (DNA and molecular machines) that are rich in CSI, and can't be
explained by material causes. This is evidence, or a record, of the work of
an intelligent designer.

Asher critiques Meyer's use of the phrase "uniform and repeated
experience" because he says that "Another 'uniform and repeated experience' that we have about intelligent agents is that they have left behind a

plethora of evidence when and wherever they have existed."[10] He claims we lack such artifacts (like "garbage" or "plastics") to give evidence of intelligent designers in the deep past on Earth.

But why plastics? Maybe the designer had advanced technologies that didn't leave behind such garbage.

Accordingly, the following simple questions show why Asher's critiques are misguided:

- Must an intelligent agent always have teeth?
- Must an intelligent agent always have bones?
- Must an intelligent agent always leave waste or garbage?
- Must an intelligent agent always leave behind an infrastructure?

If the answer to any of those questions is "no," then Asher's critique is flawed. In fact, the answer to all the questions is "No."

ID doesn't require that the intelligent designer be identical to humans—having teeth, bones, and leaving behind waste, garbage, and an infrastructure. ID simply requires that the intelligent designer be intelligent. There's no hiding the ball here: "intelligent design" means exactly what it says—intelligent design. Only by adding these additional superfluous—and might I even say unreasonable—requirements to what it means to be an "intelligent agent" can Asher critique ID.

Indeed, Asher concedes that ID argues that life arose due to "something *like* human ingenuity" (emphasis added)—not necessarily from an intelligent being identical to humans. As a result, we can detect design if we find evidence that a human-like intelligence—one that produces the high CSI and machine-like structures found throughout biology—was at work in the past. There is no need to require that the intelligence be identical to that of humans in every way (connected with bones and teeth, producing waste, and so on). What matters is whether the agent has human-like intelligence; if it does, we can potentially detect its actions.

Asher wants to be able to find something the designer left behind, and indeed there is something that intelligent agents do leave behind:

high CSI. Thus, maybe the designer *did* leave behind such evidence—i.e., in the form of high CSI in our DNA. In fact, in *Evolution and Belief*, Asher concedes that Meyer argues in this way: "Meyer argues that one such artifact has already been found. It is DNA itself... in a software-like, digital code." Asher can't accept this, however. He writes:

> While the complexity of DNA makes an interesting analogy to human creative expression, the analogy falls short as proof of human-like intelligence as the cause behind biodiversity for the philosophical, theological, and biological reasons enumerated here and elsewhere.[11]

Like all historical sciences, ID doesn't claim to provide "proof," but it does show that the best explanation for the high CSI in life is intelligence. Nonetheless, Asher is wrong to claim that the similarity between DNA and software or language is a mere analogy. As Hubert Yockey explains:

> It is important to understand that we are not reasoning by analogy. The sequence hypothesis [that the exact order of symbols records the information] applies directly to the protein and the genetic text as well as to written language and therefore the treatment is mathematically identical.[12]

Though Yockey is no ID proponent, he rightly observes that the informational properties of DNA are mathematically identical to those of language. Thus, the argument for design is much stronger than a mere appeal to analogy. It's based upon finding in nature the precise type of information that, in our experience, only comes from intelligence. This isn't a "proof" of design, but it does show that intelligent design is the best explanation for high CSI in nature.

ID isn't fundamentally opposed to looking for evidence of the designer's teeth or trash, and if we find them, then fine. But since the defining property of intelligent agents is that they produce high CSI, looking for CSI—not teeth or trash—seems like a better place to start.

Asher's article for the *Huffington Post* protests that "Meyer called my argument a 'new objection' to Intelligent Design" but "my objection to his professed methodology is not new." New or not, I can't remember anyone else who has argued that if we don't find the designer's teeth or trash, we can't detect design. While Asher thinks his argument "is not new" what surely isn't new are the rebuttals to his argument: Meyer responded to them in *Darwin's Doubt* and I responded to them at *Evolution News & Views*.[13] Asher's recent piece restates his original arguments, but fails to mention or answer any of our rebuttals.

Well, at least his objection, even if fundamentally flawed and previously answered, is phrased in a civil, serious manner. If that sounds like tepid praise, I suppose it is.

Notes

1. Robert J. Asher, "Republicans and the Unsung Fossils (Starting With 'A')," *Huffington Post*, August 28, 2012, http://www.huffingtonpost.com/robert-j-asher/republicans-intelligent-design_b_1823426.html.

2. Casey Luskin, "Huffington Post Author Invents Claims about *Explore Evolution* and Pop-Paleontology," *Evolution News & Views*, September 4, 2012, http://www.evolutionnews.org/2012/09/read_the_book_b063871.html.

3. Robert J. Asher, *Evolution and Belief: Confessions of a Religious Paleontologist* (Cambridge: Cambridge University Press, 2012), 32.

4. Asher, *Evolution and Belief*, 32.

5. Casey Luskin, "Citation Bluffs and Other 'Garbage' Arguments in *Evolution and Belief*," *Evolution News & Views*, January 14, 2013, http://www.evolutionnews.org/2013/01/citation_bluffs068241.html.

6. Casey Luskin, "Robert Asher's "Impoverished Creator" vs. Intelligent Design," *Evolution News & Views*, January 12, 2013, http://www.evolutionnews.org/2013/01/robert_ashers_i068231.html.

7. Robert J. Asher, "A New Objection to Intelligent Design," *Huffington Post*, January 9, 2014, http://www.huffingtonpost.com/robert-j-asher/a-new-objection-to-intell_b_4557876.html.

8. Ibid.

9. Asher, *Evolution and Belief*, 34.

10. Ibid.

11. Asher, *Evolution and Belief*, 35.

12. Hubert P. Yockey, "Self-Organization Origin of Life Scenarios and Information Theory," *Journal of Theoretical Biology* 91, no. 1 (July 7, 1981): 13–31.

13. See *Darwin's Doubt*, 392–398. See also Casey Luskin, "Stephen Meyer's Rebuttal to Robert Asher's 'Mechanism' Argument that ID is 'Anti-Uniformitarian'," *Evolution News & Views*, January 15, 2014, http://www.evolutionnews.org/2014/01/meyer_asher_rebuttal081161.html.

32.

BioLogos Delivers a

Raft of Reviews

David Klinghoffer

THE TEAM OF THEISTIC EVOLUTIONISTS AT BioLogos HAS HAD their ups and downs, undergoing not infrequent evolutions in their leadership staff but maintaining a consistently critical stance in relationship to the scientific theory of intelligent design. So I found it notable when I stopped by their website and saw they were planning a series of responses to *Darwin's Doubt*.[1]

This was a year and two months after the hardback appeared. It was almost three months since the paperback came out with its new Epilogue by Dr. Meyer responding to his more timely critics. There must be some backstory to explain the editorial decision at BioLogos to roll out an armada of responders to reply to *Darwin's Doubt*.

In a reflective essay that serves as an introduction to the series, current BioLogos president Deborah Haarsma promised responses from paleontologist Ralph Stearley, philosopher and historian Robert Bishop, geneticist and former BioLogos president Darrel Falk (currently BioLogos Senior Advisor for Dialogue), and theologian Alister McGrath, on top of previous remarks by BioLogos Fellow for genetics Dennis Venema.

That's a lot of writers, although, as Dr. Haarsma also indicated, Stearley's review was previously published in the journal *Perspectives on Science and Christian Faith*, while Dr. Bishop, the philosopher and historian, promised to "address the overall argument of the book, assessing the rhetorical strategies"—which sounds like something other than a scientific evaluation.

McGrath, wrote Haarsma, would not be responding to *"Darwin's Doubt* in particular, but to the overall apologetics approach of Intelligent Design." Except that intelligent design isn't a form of apologetics, but never mind.

Amid previously published material and theological and philosophical reflections on ID, it seemed, then, that the only new scientific critique of Meyer's book in this group of articles would be from Darrel Falk.

A couple of things of note leapt out from Deborah Haarsma's post. First, the acknowledgment that as Christians, the BioLogos team necessarily endorses some form of "intelligent design." As for the kind of ID that they don't accept, Dr. Haarsma capitalizes it ("Intelligent Design") and says this about the difference between her view and ours:

> The biggest difference is in how the two views counter atheistic evolutionism. Both reject the idea that the science of evolution disproves God or replaces God, but take very different approaches. Intelligent Design claims that the current scientific evidence for evolution is weak, and argues that a better explanation would make explicit reference to an intelligent designer. Evolutionary Creation claims that the current scientific evidence for evolution is strong and getting stronger, but argues that the philosophical and religious conclusions that militant atheists draw from it are unwarranted. Evolutionary creationists respond to atheists by pointing out that in Christian thought, a scientific understanding of evolution does not replace God. God governs and sustains all natural processes, from gravity to evolution, according to his purposes.[2]

It would be more accurate to write that ID says the evidence for the Darwinian evolutionary mechanism—as an explanation of the whole historical development of complex life—is weak. Beyond this, she seems to be saying that ID's response to evolutionary atheism is a scientific one, while the response from BioLogos concedes on the science and takes up its argument on "philosophical and religious" grounds alone.

And that too sounds correct. It's why I, like many other people, find the case for ID a more compelling, objective, and interesting one than arguments for theistic evolution. The same idea seems implicit as she continues:

> At BioLogos, we embrace the historical Christian faith and up-hold the authority and inspiration of the Bible. Several leaders at the Discovery Institute, including Meyer, share these commitments. The organization, however, has chosen not to make specific religious commitments, welcoming Jews, Muslims, and agnostics as well as Christians. This difference is integral to our contrasting approaches to apologetics. DI seeks to make the case for the designer in a purely scientific context, without specifying who the designer is. At BioLogos, we take the approach that science is not equipped to provide a full Christian apologetic. Rather, we believe in the triune God for the same reasons most believers do—because of the evidence in the Bible, personal spiritual experience, and recognition that we are sinners who need the saving work of Jesus Christ.[3]

The reference to "several leaders" at Discovery Institute sharing her Christian "commitments" sounds like a dig at our Christian bona fides, which as a Jew, I don't much mind. The Young Earth Creationists at Answers in Genesis have the same complaint about us, though it's more plainly expressed.

The important point, as she puts it, is that ID advocates "seek to make the case for the designer in a purely scientific context." And that, again, is true. It is our distinction. It is what makes arguments for intelligent design such an important phenomenon in science and culture, with roots stretching from Athens and Jerusalem to Maimonides and Aquinas, from the scientific revolution to the cutting edge of biology and cosmology today.

It would also seem to make philosophical, religious, historical, or apologetic objections to ID less relevant than scientific ones. So that lone review by Dr. Falk must carry a lot of the burden for his colleagues.

Notes

1. Deborah Haarsma, "Reviewing *Darwin's Doubt*: Introduction," *The BioLogos Forum*, August 25, 2014, http://biologos.org/blog/reviewing-darwins-doubt-introduction.

2. Ibid.

3. Ibid.

33.

RALPH STEARLEY'S "WELL, MAYBE, WHO KNOWS?" REVIEW

Paul Nelson

MORE THAN 18 MONTHS AFTER ITS PUBLICATION, *DARWIN'S Doubt* continued to stir discussion and debate, but that discussion all too often has savored of a peculiar and unsatisfying incompleteness. As an observer of the debate, I often wondered if the critics read the same volume that I did. Thus, the organization BioLogos commenced a multi-part response, which I began reading with high hopes of finding the reviewers actually grappling with Stephen Meyer's central theses.

The first part of the BioLogos response[1] to Meyer's book consists of a blog post recommending Calvin College paleontologist Ralph Stearley's December 2013 essay review about the Cambrian explosion,[2] published in *Perspectives on Science and Christian Faith*, the last part of which considered *Darwin's Doubt* (in addition to two other books on the topic). Stearley reprises worries put forward earlier by other critics—about such matters as the exact timing of the Cambrian explosion, the "small shelly" fauna, and Cambrian ecologies—but his bottom line is so ambivalent that it is impossible to say if he agrees with Stephen Meyer or not. He certainly provides no scientific refutation of Meyer's main scientific arguments. Instead he actually acknowledges the inadequacy of the neo-Darwinian mechanism as an explanation for major innovation in the history of life. If Stearley had stepped into a voting booth, we would find him still there, with the curtain drawn, deliberating over his choices. I'll comment below on why he can't decide. It's a philosophical, not a scientific, dilemma.

First, however, let's dispense with the peripheral issues, all of which have been previously addressed by Stephen Meyer or others. Stearley disputes what he calls Meyer's "minimalist interpretation" of the length of the Cambrian explosion, saying that, by ignoring the appearance of the "small shelly" fauna, *Darwin's Doubt* exaggerates the abruptness of the event. But as Casey Luskin points out, Meyer did not ignore these fossils.[3] Moreover, as Meyer himself explains, even expanding the geological interval (from 10 to 25 million years, or more) does little to solve the relevant problems of new information and anatomical innovation.[4] It's a bit like arguing about the length of a bank robbery: twenty minutes, three hours, all night? In the morning, the vault is still empty. Someone did it. The vault did not empty itself.

Nor are the ecologies of the early- to mid-Cambrian the issue of central interest. Stearley claims that "new adaptive niches" opened during the Cambrian. That may well be true, but environmental changes are hardly *sufficient* to cause the origin of the wide array of novel animal body plans. Paleontologist Douglas Erwin and colleagues identify this confusion of necessary and sufficient conditions. Raising the oxygen level of pre-Cambrian oceans, for instance, may have allowed oxygen-fueled animal metabolism, required for elaborate body plans, to flourish, but greater amounts of oxygen alone could never have caused the same complexity. As Erwin et al. explain, "a permissive environment does not explain innovations in metazoan architecture."[5] Put a few yeast cells in an enormous chemostat, with nutrients, and slowly increase the oxygen levels, over a really, *really* long time. Open the chemostat. Trillions of yeast cells; no animals.

Stearley also complains about Meyer's treatment of alternative evolutionary theories, such as those proposed by developmental biologist

Eric Davidson or self-organization theorist Stuart Kauffman. Here at last we see some glimmer of the real issue, namely, what message may we take away from the ongoing failures of materialistic theories to solve the problem of the Cambrian explosion? Or, put another way—*if the signal of nature appears to indicate intelligent design, may we follow that signal where it leads?* Or are we constrained to seek a materialist solution, come what may?

Stearley acknowledges that many leading evolutionary theorists are deeply unhappy with the received neo-Darwinian account for major events in the history of life, such as the Cambrian explosion. Yet their unhappiness is still not enough, he argues, to move them out of the City of Materialism:

> while it is true that Goodwin and others believe that their discoveries pose a major challenge to neo-Darwinian orthodoxy, this does not cause them to abandon their belief that the history of life can be explained as the outcome of biological processes![6]

One shouldn't make too much of a punctuation mark, I suppose, but Stearley's exclamation point at the end of that passage, reinforcing the non-negotiability of materialism for evolutionary theory, is telling, especially when juxtaposed against his own ambivalence about the possibility of detecting design as a genuine empirical finding. "I admit that, by temperament," he writes, "I am inclined to see design in nature, and so I resonate with some of Meyer's arguments."

Resonate?

Design, if actual, is no more a question of "inclining to see," "temperament," or "resonating," than the atomic number of an element. The evidence and arguments compiled in *Darwin's Doubt* are not menu suggestions about—if mood happens to strike you, on any given day—preferring one breakfast jam over another. If real, design is a datum of nature, like it or not. Bad philosophies of science, like materialism, need to get out of the way.

And unfortunately Stearley seems not to know his own mind on this point. Although he says he is "temperamentally inclined" to see design, he pushes that inference away with his other hand. "I am not sure," he concludes, "that it is our place to know [about intelligent design]. If that is so, perhaps our efforts to obtain certainty in seeing his design will end in frustration." Stearley cannot decide if materialism—or, methodological naturalism, to give the doctrine its domesticated name—governs science, or if design is truly detectable.

In the long run, epistemic ambivalence like that, vacillating on the horns of a philosophical dilemma, will prove sterile, if not indeed deadly, because ambivalence robs evidence of its power to yield knowledge. When one says "Sorry, but I cannot know **X**," then it simply does not matter how powerful or compelling the evidence for **X** may be. An *a priori* move has destroyed what should be a ready inference.

Strange asymmetry: materialist evolutionary theory may pursue its investigations, with the promise of genuine discoveries awaiting, whereas intelligent design necessarily lies beyond the horizon of knowledge in a mist of uncertainty. Those who prefer their yes to be yes (and no to be no) will never settle for this asymmetrical playing field. Science is hard, and inferences are tricky, but your blood really does circulate—and design, if it's out there to be detected, awaits our hard work, and will reward us when we find it out.

Temperament? Ah, don't worry about that. Not relevant.

Notes

1. Ralph Stearley, "Reviewing *Darwin's Doubt*," *The BioLogos Forum*, August 26, 2014, http://biologos.org/blog/reviewing-darwins-doubt-ralph-stearley.

2. Ralph Stearley, "The Cambrian Explosion: How Much Bang for the Buck?," *Perspectives on Science and Christian Faith* 65 no. 4 (2013): 245–257, http://www.asa3.org/ASA/PSCF/2013/PSCF12-13Stearley.pdf.

3. Casey Luskin, "Small Shelly Fossils, and the Length of the Cambrian Explosion," *Evolution News & Views*, October 23, 2013, http://www.evolutionnews.org/2013/10/small_shelly_fo078261.html.

4. Stephen C. Meyer, "More on Small Shelly Fossils and the Length of the Cambrian Explosion: A Concluding Response to Charles Marshall," *Evolution News & Views*, October 23, 2013, http://www.evolutionnews.org/2013/10/more_on_small_s078251.html.

5. Douglas Erwin et al., "The Cambrian Conundrum: Early Divergence and Later Ecological Success in the Early History of Animals," *Science* 334 (November 25, 2011): 1091–1097, http://www.sciencemag.org/content/334/6059/1091.full.html.

6. Stearley, "The Cambrian Explosion."

34.

STILL AWAITING ENGAGEMENT

Paul Nelson

BIOLOGOS SOON POSTED THE NEXT SEGMENT OF ITS COMPREHEN-
sive response to *Darwin's Doubt*. Disappointingly, these entries—
Part 1[1] and Part 2[2] of philosopher of science Robert Bishop's four-part
critique—did not reply to the scientific arguments or evidence presented
in Meyer's book. Instead, Bishop focuses on what he calls the "rhetorical strategy" of *Darwin's Doubt* and its framing of the current status of
evolutionary theory. Bishop finds fault both with the rhetoric and the
framing of the book, but he does so by mischaracterizing Meyer's presentation of evolution. Moreover, Bishop's critique contains serious errors in its discussion and understanding of evolutionary theory, which
vitiate his case.

Does Meyer Employ a "Divide-and-Conquer" Strategy?

IN *DARWIN'S DOUBT*, Meyer argues that intelligent design best explains
the origin of the biological information necessary to build the animals
that appeared abruptly in the Cambrian period. In support of this argument, Meyer demonstrates that neither textbook neo-Darwinism nor
more recent versions of evolutionary theory provide an adequate explanation for the explosion of novel biological form and information (both
genetic and epigenetic) that arose in the Cambrian period.

In Chapters 8–14, he makes several separate evidentially based arguments to demonstrate the inadequacy of the neo-Darwinian natural selection/random mutation mechanism as an explanation for the origin of animal life. But in Chapters 15 and 16, he also explores the ideas of a wide range of evolutionary biologists who have expressed dissatisfaction with current neo-Darwinian theory and formulated alternative evolutionary models. Meyer then critiques these alternative proposals as well, showing in each case that they either fail to address the problem of the origin of the necessary biological information or that they simply presuppose earlier unexplained sources of such information.

Bishop finds this approach "misleading," because—he argues—the alternatives that Meyer addresses are not genuinely replacements for current neo-Darwinian theory, but are merely additions or expansions to a basically sound core. He calls Meyer's analysis a "divide-and-conquer" strategy:

> Meyer rightly points out that there has been a long history of trying to understand the details of macroevolutionary change in neo-Darwinian evolution... Meyer successively reviews a variety of attempts, such as evo-devo [evolutionary developmental biology] to rectify this shortcoming in macroevolution. Each attempt surveyed is presented to the reader as being in competition with and a replacement for neo-Darwinian evolution (population genetics and natural selection)... [but] researchers working in evo-devo typically don't see themselves as replacing population genetics and natural selection.[3]

Well, evo-devo researchers and many others are seeking to replace something that they perceive as wrong with textbook theory. More often than not, that something is one, or indeed more than one, of neo-Darwinism's key pillars: (1) small-scale, randomly arising variations and mutations as the raw materials of evolution; (2) natural selection as the primary creative process; and (3) heritability grounded in the vertical transmission of DNA.

Contrary to Bishop's critique, Meyer is careful to spell out in each case exactly which pillar is under attack by evolutionary theorists seeking alternatives, and why the proposed alternatives themselves are nonetheless unable to explain the Cambrian explosion. And Meyer correctly points out (again, *contra* Bishop) that the proposed alternatives and textbook theory are mutually contradictory. The new proposals cannot be seamlessly grafted onto a core of existing theory, because the investigators in question see existing theory as gravely defective, not basically sound. To interpret the situation otherwise would be at best naïve.

Let's consider some examples. Geneticist Michael Lynch of Indiana University, whose alternative evolutionary proposals are discussed extensively in Chapter 16 of *Darwin's Doubt*, has argued that "nothing in biology makes sense except in the light of population genetics."[4] At first blush, this statement appears to support Bishop's claim that proposed alternatives build on, but do not replace, existing theory. After all, population genetics and natural selection go hand-in-hand, right?

But watch what Lynch does with his dictum about population genetics. He uses it as a battering ram to smash right through the doors of textbook (i.e., received) theory. Starting from the non-negotiable principles of population genetics, Lynch asserts, one must acknowledge that non-adaptive causes such as random genetic drift.

> dictate what natural selection can and cannot do. Although this basic principle has been known for some time, it is quite remarkable that most biologists continue to interpret nearly every aspect of biodiversity as an outcome of adaptive processes. This blind acceptance of natural selection as the only force relevant to evolution has led to a lot of sloppy thinking, and is probably the main reason why evolution is viewed as a soft science by much of society.[5]

All right—so what do the textbooks say? What is the core neo-Darwinian theory, which Bishop claims is basically sound?

We can consult Dobzhansky, Ayala, Stebbins, and Valentine in a standard and widely used text: "According to the theory of evolution... natural selection is the process responsible for the adaptations of organisms, and also the main process by which evolutionary change comes about."[6] Or consider George Williams's classic, influential analysis of natural selection: the process, he argues, provides "the only acceptable theory of the genesis of adaptation."[7] Or Dawkins: "Adaptation cannot be produced by random drift, or by any other realistic evolutionary force that we know of save natural selection."[8]

Looks like a fundamental theoretical conflict, doesn't it? In *direct opposition* to received theory, Lynch wants to dump natural selection from its central explanatory role and isn't coy about stating his reasoning or motivation: "[I]t is a leap to assume that selection accounts for all evolutionary change, particularly at the molecular and cellular levels. The blind worship of natural selection is not evolutionary biology. It is arguably not even science."[9]

"Not even science" is hardly the sort of language one expects from a biologist mildly or moderately discontent with current theory, looking to graft his additional considerations onto a more or less healthy core theory. The same is the case with other biologists seeking alternatives, whom Bishop wants to tuck into the neo-Darwinian fold.

Consider, for instance, Caltech developmental biologist Eric Davidson, whom Bishop says (in the online supplement to his review) is "working out a synthesis of evolutionary development and neo-Darwinian evolution." Really?

In the opening of a 2011 paper that Bishop himself cites, Davidson gives his candid assessment of neo-Darwinian theory:

> it gives rise to *lethal errors* in respect to evolutionary process. Neo-Darwinian evolution is uniformitarian in that it assumes that all process works the same way, so that evolution of enzymes or flower colors can be used as current proxies for study of evolution of the body plan. It *erroneously assumes* that change in protein coding sequence is the basic cause of change in developmental program; and it *erroneously assumes* that evolutionary change in body plan morphology occurs by a continuous process. *All of these assumptions are basically counterfactual.* This cannot be surprising, since *the neo-Darwinian synthesis from which these ideas stem was a pre-molecular biology concoction focused on population genetics and adaptation natural history, neither of which have any direct mechanistic import* for the genomic regulatory systems that drive embryonic development of the body plan.[10]

The emphases are mine, but they're probably unnecessary—it is difficult to miss Davidson's thrust: As far as the origin of animal body plans is concerned, neo-Darwinism isn't incomplete or insufficient. It is dead wrong. And no biologist in his good senses would seek to synthesize his ideas with a corpse.

We could go on in this vein for pages, but the point is clear. When a scientist says that something is wrong with a current theory, we need to pay attention to the details of his objection. Is he troubled by some minor matter or speaking about the core of the theory? The alternative evolutionary proposals presented and critiqued in *Darwin's Doubt* reject core propositions of neo-Darwinian theory, not peripheral theoretical commitments. Stephen Meyer is not dividing and conquering, but simply reporting. It's up to the reader to decide if he wants to stay with neo-Darwinism or try his luck elsewhere.

Does Meyer Shift the Questions Arising in Evolutionary Theory?

Bishop's second critique of *Darwin's Doubt* turns on what he calls the book's "question-shifting strategy." As he puts it:

> This strategy involves equivocating on the notion of origin. In the biology and paleontology literature, when scientists discuss the origin of Cambrian body plans, they mean *the modification and diversification of body plans from preexisting body plans.*[11]

The statement in italics (Bishop's own emphasis) is simply false. I want to add *inexplicably* false, because—with a moment's reflection—one realizes that even a "preexisting" body plan must be a body plan for *some* kind of animal and must have arisen at some discrete interval in Earth history. Consider: one billion years ago, no animals—500 million years ago, lots of animals of many different types.

No matter how one carves up the puzzle, one cannot bracket the problem of primary origins indefinitely. In other words, the problem of the Cambrian explosion *by definition* includes the origin of the first animals, meaning the first body plans (whether those plans diversified later or not). Given any evolutionary approach to the data, there is no way around answering the question, "How did the first animals come to be?"

Bishop makes this serious error more inexplicable by conflating the problem of the origin of animal body plans with the problem of the origin of life, and with Darwin's treatment of that problem in the *Origin of Species*:

> [Modification and diversification] is the customary usage in the literature since Darwin's publication... where he makes clear that he is seeking only to explain speciation, not how the first species arose. The latter question is the origin of life issue, a separate question from how an ancestor species may be connected with descendant species through descent with modification.[12]

But the origin of metazoan multicellularity and the diverse macroscopic architectures of the Cambrian explosion are chapters in the evolutionary narrative *hundreds of pages later* than the origin of life itself. Indeed, the origin of life and the origin of animals constitute discrete events in the history of life separated by billions of years of Earth history. Along the way, we must pass through such earlier, but absolutely necessary, chapters as the division of the three primary domains (Bacteria, Archaea, Eukarya), the origin of cell organelles, the origin of eukaryotic cytoskeletal complexity, the origin of colonial protists, the origin of sexual reproduction, cellular differentiation, developmental pathways, and so forth—that is, through a whole lot of complexity-building and evolutionary *origins* events. Throughout the narrative leading to animal body plans, tens of thousands of novel biological traits must come to be where they did not exist before—namely, *they must originate*.

That is the "customary usage" since Darwin, unless one tries to offload the hard problems by simply naming them "origins" questions and stipulating that evolutionary theory does not address them. But then what? If the theory of evolution is anything, it is an attempt to explain how X came to be—i.e., *originated*—where X did not exist before. And X includes the first animal (metazoan).

This problem has nothing whatsoever to do with the origin of life, or only the most tenuous connection. Indeed, two different branches of evolutionary theory—chemical and biological evolutionary theory—address these two separate questions. For all that, one might assume (as Darwin seemed to intimate from time to time) that the first cell was divinely created, and the puzzle of animal origins would still remain. Thus, when Bishop writes that "the logical fallacy... is Meyer's falling into equivocation on two different senses of 'origin' and shifting all diversification questions to origin of life questions," he is just flatly mistaken.

Still Awaiting Engagement

THUS, AT the end of the day, it really doesn't matter whether the contemporary evolutionary theorists that Meyer discusses in *Darwin's Doubt* are attempting to supplement neo-Darwinian theory, replace it with something fundamentally new, or replace some, but not all, parts of the theory. What matters is whether any of these theories can explain what needs to be explained: the origin of novel animal body plans and the biological information necessary to produce them.

In *Darwin's Doubt*, Meyer argues that neither neo-Darwinism, nor recently proposed alternative theories of evolution (punctuated equilibrium, self-organization, Lynch's neutral theory, neo-Lamarckian epigenetic inheritance, evolutionary developmental biology, and natural genetic engineering) have solved this problem. And certainly Bishop himself offers no solution to it. Indeed, by focusing his analysis on the alleged rhetorical strategy of the book rather than its scientific case, Bishop fails to address Meyer's central arguments—a failed rhetorical strategy if ever there was one.

Notes

1. Robert C. Bishop, "The Extended Synthesis (Reviewing *Darwin's Doubt*: Robert Bishop, Part 1)," *The BioLogos Forum*, September 1, 2014, http://biologos.org/blog/the-grand-synthesis-reviewing-darwins-doubt-robert-bishop-part-1.

2. Robert C. Bishop, "The Extended Synthesis (Reviewing *Darwin's Doubt*: Robert Bishop, Part 2)," *The BioLogos Forum*, September 2, 2014, http://biologos.org/blog/two-rhetorical-strategies-reviewing-darwins-doubt-robert-bishop-part-2.

3. Ibid.

4. Michael Lynch, "The frailty of adaptive hypotheses for the origins of organismal complexity," *Proceedings of the National Academy of Sciences, USA* 104 (2007): 8597, http://www.pnas.org/content/104/suppl_1/8597.full.

5. Michael Lynch, *The Origins of Genome Architecture* (Sunderland, MA: Sinauer Associates, 2007), xiii.

6. T. Dobzhansky, F. Ayala, G. Stebbins, and J. Valentine, *Evolution* (San Francisco: W. H. Freeman, 1977), 504.

7. George Williams, *Adaptation and Natural Selection* (Princeton: Princeton University Press, 1966), 251.

8. Richard Dawkins, *The Extended Phenotype* (San Francisco: W. H. Freeman, 1982), 19.

9. Lynch, *The Origins of Genome Architecture*, 369.

10. Eric Davidson, "Evolutionary bioscience as regulatory systems biology," *Developmental Biology* 357 (2011): 35–36, http://www.sciencedirect.com/science/article/pii/S0012160611000911. Emphasis added.

11. Bishop, "The Extended Synthesis," Part 2.

12. Ibid.

35.

ARE BIOLOGISTS COMING

TO REJECT NEO-DARWINIAN

EVOLUTION?

Casey Luskin

A S PAUL NELSON OBSERVES IN THE PRECEDING CHAPTER, ROBERT Bishop's critique of *Darwin's Doubt* for BioLogos denies that evolutionary biologists are entering a post-Darwinian world and abandoning neo-Darwinism. Dr. Bishop writes:

> Meyer successively reviews a variety of attempts, such as evo-devo [evolutionary developmental biology] to rectify this shortcoming in macroevolution. Each attempt surveyed is presented to the reader as being in competition with and a replacement for neo-Darwinian evolution (population genetics and natural selection)… [but] researchers working in evo-devo typically don't see themselves as replacing population genetics and natural selection.[1]

Of course none of these post-Darwinian biologists says that natural selection doesn't occur. (Advocates of intelligent design don't say that, either.) Rather, they say that certain forces—natural selection or random mutation, for example—might not be as important in driving and shaping evolution as was once thought. You could, as Stephen Meyer does, cite many examples of evolutionary biologists specifically critiquing the core tenets of neo-Darwinism. (See Chapters 15 and 16 of *Darwin's Doubt*.)

Many biologists say they seek new mechanisms of macroevolution—thus implicitly rejecting the neo-Darwinian paradigm. Even more interesting is that, in explaining their doubts about Darwinian theory and their interest in finding an alternative, many point to the Cambrian explosion. Consider how Oxford's Simon Conway Morris has described current thinking in biology:

> The other day I was walking past an immense building from which emanated an uproar. The noise was terrific. A door swung open and, looking in, I saw to my surprise that nearly everyone was dressed in white. But, strange to say, there was not one pulpit but two. The crowd surged back and forth, spotless laboratory coats streaming in the rush. From one pulpit the proclamation rang out: *"The Cambrian 'explosion' is real!!! Hundreds of phyla evolved, almost instantaneously. Listen, neo-Darwinism is in terminal crisis, we must summon forth new mechanisms of macroevolution."* From the other pulpit, however, I heard the following: "No, the Cambrian 'explosion' is a mirage, a mere artefact! For aeons tiny animalcules slithered through the slime, avoiding fossilization, hoarding their Hox genes, swaying to the sonorous tick of molecular horology."[2]

According to Conway Morris, the two main schools of thought about the Cambrian explosion either (1) claim it is simply an illusion, and an artifact of an imperfect fossil record, or (2) seek "new mechanisms of macroevolution," rejecting neo-Darwinian explanations because they are "in terminal crisis." According to this eminent scholar, many biologists question the neo-Darwinian paradigm itself. They have turned to what Conway Morris elsewhere calls "post-Darwinian" models of evolution.

Whether these post-Darwinian models solve the mystery of the Cambrian explosion (and the origin of new body plans) is another question. As Meyer reviews the available alternatives in Chapters 15 and 16 of *Darwin's Doubt*, they don't look very promising. But the fact of the matter is that biology continues to uncover evidence unanticipated by Darwin or by the subsequent neo-Darwinian paradigm. This evidence provides the impetus for post-Darwinian models of evolution.

Notes

1. Robert C. Bishop, "The Extended Synthesis (Reviewing *Darwin's Doubt*: Robert Bishop, Part 2)," *The BioLogos Forum*, September 2, 2014, http://biologos.org/blog/two-rhetorical-strategies-reviewing-darwins-doubt-robert-bishop-part-2.

2. Simon Conway Morris, "Nipping the Cambrian 'explosion' in the bud?," *BioEssays* 22 (2000): 1053–1056. Emphasis added.

36.

LEADING THEISTIC EVOLUTIONIST

PRAISES *DARWIN'S DOUBT*

David Klinghoffer

A S WE'VE ALREADY EXPLAINED, BioLogos IS A CHRISTIAN ORGA-
nization founded by geneticist Francis Collins, current director of
the National Institutes of Health (NIH). In years past the group has
emphasized what it regards as the necessity of embracing Darwinian
evolution over the competing theory of intelligent design. It has broad-
cast this message to the Christian community with some success.

Meanwhile, though, much to the credit of our friends at BioLogos,
leading spokesmen for the theistic evolutionary view have sought to engage in constructive dialogue with advocates of ID.

It's gratifying to report that our conversations have borne fruit. At
the BioLogos website, developmental geneticist and past BioLogos president Darrel Falk has contributed a two-part review of *Darwin's Doubt*
that is, candidly, remarkable.[1, 2]

For Dr. Falk, now Senior Advisor for Dialogue at BioLogos, *Darwin's Doubt* is "amazingly effective" and "Meyer has successfully put his
finger on one of the great mysteries in evolutionary biology today."

Regarding the scientific evidence of design in nature, it would certainly be too much to read the review as an indication that theistic evolu-

tionists are close to changing their fundamental view. Indeed Dr. Falk's articles come in the context of a series of critical articles on the BioLogos site responding to Meyer. (See Paul Nelson's and Casey Luskin's comments in Chapters 33 to 35.)

Falk affirms the enigma of the Cambrian explosion with its "rapid" production of new body plans from no identifiable predecessors, just as Meyer explains it in *Darwin's Doubt*. Echoing Meyer, he confirms that current evolutionary thinking is frustrated in seeking to explain how complex animal life arose. Writes Falk:

> The big mystery associated with the Cambrian explosion is the rapid generation of body plans *de novo*. There was never a time like it before, nor has there ever been a time like it again since. Stephen is right about that. Also, as he points out, the big question in exploring the generation of new body plans in that era is how this squares with the resistance of today's gene regulatory networks to mutational perturbation (i.e. they seem to be almost impossible to change through genetic mutation because virtually all such alterations are lethal). We really have little idea at this point how things would have worked to generate body plans *de novo* back then given the sensitivity of the networks to perturbation today.[3]

More:

> The depth of knowledge [Meyer] displays in molecular genetics, developmental biology, and population genetics in addition to paleontology, animal diversity, biochemistry, and even some cell biology is very impressive.[4]

In inferring design, however, Falk writes, "I think [Meyer is] wrong, of course." And:

> So have I softened on Intelligent Design as a scientific endeavor? I don't think so, but I have grown to appreciate the skill and the sincerity of various individuals I have met in the ID movement over the last five years.[5]

Nevertheless, for Dr. Falk at least, *Darwin's Doubt* is a breakthrough. While still rejecting the evidence for design presented in previous books about ID by Meyer, Behe, and others, he warmly praises the scientific argument in the book, identifies no fault in its presentation of the relevant science, and, significantly, takes issue with his colleague Robert Bishop's denial that biologists are having second thoughts about Darwinian theory.

In response to Bishop, Falk asks, "Does Stephen Meyer exaggerate the nature of the rethinking going on in mainstream evolutionary developmental biology?" He answers:

> I don't think so. Many evolutionary developmental biologists think that we are on the verge of a significant re-organization in our thinking about the mechanics of macroevolution. The much respected developmental biologist Scott Gilbert states: "If the population genetics model of evolutionary biology isn't revised by developmental genetics, it will be as relevant to biology as Newtonian physics is to current physics."[6]

> That and many other similar statements that I've seen in the literature[7] really do suggest that we are on the cusp of some major rethinking about the forces at work in macroevolution. Those studies will focus more on how biological information is generated, changed, and used, and less on the natural selection filter.[8]

In short, he confirms the negative side of Meyer's argument. Given that Dr. Falk is the only biologist among the lineup of reviewers offered by BioLogos, that is important.

What about the positive argument, for ID itself? Falk's review is helpful in clarifying where the conversation about intelligent design and theistic evolution can go from here. His real objection to design theory isn't scientific but rather a matter of philosophical sensibility or predisposition:

> [ID proponents] think the philosophical naturalism of many leading scientists has significantly influenced their conclusions, and I certainly agree that there have been times when that is the case. However, where we don't agree is that the whole applecart of evolutionary biology needs to be turned upside down and replaced with a new science—one grounded in the scientific demonstration of Intelligence. I see no scientific, biblical, or theological reason to expect that. Natural processes are a manifestation of God's ongoing presence in the universe.[9]

It seems to be precisely his *a priori* commitment to methodological naturalism that holds him back from joining us in recognizing design in biology:

> Stephen is right, that none of the other [evolutionary] models fit the bill in a fully satisfactory manner yet, but it's pretty early to declare one to be the winner on the basis of an analogy to human-designed information systems.[10]

He regards that recognition as premature, but for the methodological naturalist it must, by definition, always be premature to affirm intelligent design:

> I agree with Robert [Bishop] that it is quite a stretch to jump from the "failure" of materialistic explanations of the Cambrian explosion (so far) to a scientifically based conclusion that life is intelligently designed.[11]

Between theistic evolutionists and proponents of intelligent design there remains a sharp divide, specifically in our views of what constitutes a potentially legitimate scientific inference and what does not. So we have a lot to talk about with our colleagues at BioLogos. Falk's review does a service by making clear what needs to be on the agenda of future discussions.

Dr. Falk has good questions, too, about the future of ID as a research program. He would like to know, beyond the critique of Darwinism, how we envision intelligent design as a positive paradigm for the scientific investigation of nature. He can't picture what that would look like. He also asks what predictions ID might make. How can the theory be tested?

> How will they move forward by building a positive research program rather than a negative one based upon the critique of mainstream ideas? What are the biological predictions that will emerge from within their paradigm and how will they test them?[12]

These are wonderful challenges, and we have given them much consideration ourselves. As food for thought, I commend to Falk our reporting at *Evolution News & Views* on how mainstream scientists, without explicitly recognizing it, are already employing assumptions about design to advance biology. See, for example, Casey Luskin's coverage on the field of systems biology:

+ "When Biologists Think Like Engineers: How the Burgeoning Field of Systems Biology Supports Intelligent Design"[13]

+ "Peer-Reviewed Science: What the Field of Systems Biology Can Tell Us About Intelligent Design"[14]

Other relevant articles include:

+ "Does Intelligent Design Help Science Generate New Knowledge?"[15]

+ "How Do We Know Intelligent Design Is a Scientific 'Theory'?"[16]

+ "Intelligent design (ID) has scientific merit because it uses the scientific method to make its claims and infers design by testing its positive predictions"[17]

+ "The Positive Case for Intelligent Design"[18]

+ "A Positive, Testable Case for Intelligent Design"[19]

+ "Does intelligent design theory implement the scientific method?"[20]

And see the Appendix in *Signature in the Cell* in which Stephen Meyer lists a slew of predictions of the theory of ID.

I don't want to overstate how close Darrel Falk has drawn to us. Does he, for example, fall into old habits in discussing top ID theorists? Yes, he does. He commends Dr. Meyer, for instance, for his "sincerity" while also calling the book "somewhat of a masterpiece for accomplishing their agenda."

That seems out of place in weighing a serious and popular work like *Darwin's Doubt*, a book by a Cambridge University-trained philosopher of science and leading advocate of a scientific theory that is a force to be reckoned with in scholarly thinking, a book that is at this writing No. 1 on Amazon's Paleontology bestseller list ahead of books by Nicholas Wade, Stephen Jay Gould, and Douglas Erwin and James Valentine.

But so what? The good news in Darrel's review is very good indeed. For every Darrel Falk in the theistic evolution community, whose name we know, there are undoubtedly many others whose names we don't know who are similarly ready to break with old attitudes.

Dr. Falk deserves praise for his fresh approach to the issues at hand, and, not least, for his willingness to publicly correct a colleague, Dr. Bishop, whom he knows to be mistaken on an important matter. That takes courage.

Notes

1. Darrel Falk, "Thoughts on *Darwin's Doubt* (Reviewing *Darwin's Doubt*: Darrel Falk, Part 1)," *The BioLogos Forum*, September 9, 2014, http://biologos.org/blog/thoughts-on-darwins-doubt-reviewing-darwins-doubt-darrel-falk-part-1.

2. Darrel Falk, "Further Thoughts on *Darwin's Doubt* after Reading Bishop's Review (Reviewing *Darwin's Doubt*: Darrel Falk, Part 2)," *The BioLogos Forum*, September 11, 2014, http://biologos.org/blog/thoughts-on-darwins-doubt-reviewing-darwins-doubt-darrel-falk-part-2.

3. Ibid.

4. Falk, "Thoughts on *Darwin's Doubt*."

5. Ibid.

6. His quote is from: Suzan Mazur, "Scott Gilbert: Evolutionary Mechanisms Knish," *Scoop*, February 18, 2009, http://www.suzanmazur.com/?p=4.

7. In a footnote, Falk adduces "books or articles by: Gunter Wagner of Yale, Douglas Erwin of the American Museum of Natural History, and Marc Kirschner and John Gerhart of Harvard and the University of California, Berkeley."

8. Falk, "Further Thoughts on *Darwin's Doubt*."

9. Falk, "Thoughts on *Darwin's Doubt*."

10. Falk, "Further Thoughts on *Darwin's Doubt*."

11. Ibid.

12. Ibid.

13. Casey Luskin, "When Biologists Think Like Engineers: How the Burgeoning Field of Systems Biology Supports Intelligent Design," *Evolution News & Views*, July 17, 2014, http://www.evolutionnews.org/2014/07/when_biologists087871.html.

14. Casey Luskin, "Peer-Reviewed Science: What the Field of Systems Biology Can Tell Us About Intelligent Design," *Evolution News & Views*, July 18, 2014, http://www.evolutionnews.org/2014/07/peer-reviewed_s_3087881.html.

15. Casey Luskin, "Does Intelligent Design Help Science Generate New Knowledge?," *Evolution News & Views*, November 23, 2013, http://www.evolutionnews.org/2010/11/does_intelligent_design_help_s040781.html.

16. Casey Luskin, "How Do We Know Intelligent Design Is a Scientific 'Theory'?," *Evolution News & Views*, October 14, 2014, http://www.evolutionnews.org/2011/10/how_do_we_know_intelligent_des051841.html.

17. Casey Luskin, "Intelligent design (ID) has scientific merit because it uses the scientific method to make its claims and infers design by testing its positive predictions," *Discovery Institute*, September 8, 2008, http://www.discovery.org/a/7051.

18. Casey Luskin, "The Positive Case for Design," *Discovery Institute*, http://www.discovery.org/f/986.

19. Casey Luskin, "A Positive, Testable Case for Intelligent Design," *Evolution News & Views*, March 30, 2011, http://www.evolutionnews.org/2011/03/a_closer_look_at_one_scientist045311.html.

20. Casey Luskin, "FAQ: Does intelligent design theory implement the scientific method?," *IDEA Center*, http://www.ideacenter.org/contentmgr/showdetails.php/id/1154.

37.

METHODOLOGICAL NATURALISM:

A RULE THAT NO ONE

NEEDS OR OBEYS

Paul Nelson

A S I SAID IN CHAPTER 33, BIOLOGOS POSTED ROBERT BISHOP's multipart review of *Darwin's Doubt*, concluding with Part 3[1] and Part 4.[2] Although the discussion at BioLogos by this point was winding down, Bishop's review raised issues calling for a reply. The underlying premises of his position are shared by large numbers of science and philosophy faculty at both secular and religious universities. It is likely that, left unchallenged and unexamined, these faulty premises will continue to influence the debate.

One issue in particular deserves extended comment: the standing of methodological naturalism (hereafter, MN) as a rule for scientific inquiry. Evaluating the role of MN can make for distinctly odd bedfellows. One finds theists—such as Bishop, or Princeton philosopher Hans Halvorson—arguing in favor of the doctrine,[3] and one finds atheists and agnostics—such as Maarten Boudry and colleagues,[4] as well as Sahotra Sarkar[5] and Bradley Monton[6]—arguing against it. Bishop's use of MN in his critique of *Darwin's Doubt* reveals the enormous distance between his position and that of ID theorists. Although the distance is great, communication across the divide is still possible. We hope that our response helps to clarify the ID standpoint and indicate how a more thorough analysis of MN can help in the future.

What Methodological Naturalism Is Not (and Never Was)

IN DARWIN'S *Doubt*, Stephen Meyer argues that inferences to intelligent causation, while fully warranted by the evidence of the Cambrian explosion, run afoul of the dictum of methodological naturalism (MN). As Meyer defines MN:

> scientists should accept as a working assumption that all features of the natural world can be explained by material causes without recourse to purposive intelligence, mind, or conscious agency.[7]

As Meyer later explains, the fatal defect in MN is not hard to find: "if researchers refuse as a matter of principle [namely, MN] to consider the design hypothesis, they will obviously miss any evidence that happens to support it."[8] One cannot evaluate the evidence for or against any hypothesis that has been ruled out *a priori*. For this and other reasons, ID theorists regard MN as an obstacle to knowledge and hence a methodological rule that we would be better off without.

Bishop cannot see the harm in MN. Quite the reverse; in his view, "methodological naturalism is the way scientific investigation has been done since before the time of the Scientific Revolution."[9] The rule of MN—a reasonable and philosophically neutral boundary, as he sees it—simply represents an approach to scientific investigation that seeks to "take the biological phenomena on their own terms to understand them as they actually are."[10]

Now, who could disagree with that cheerful formulation of MN? No one, really—certainly not Steve Meyer or any other ID theorist. Consider: if the phenomena of the Cambrian explosion in fact implicate intelligent design, then of course we should try to explain those events, to employ Bishop's phrase, "on their own terms... as they actually are."[11]

Expressed that way, ID and MN would be entirely congruent, and you wouldn't be reading this chapter.

But, as the rest of his review makes clear, that's not at all what Bishop means by MN. Rather, Bishop means that *phenomena are to be understood and explained solely via material or physical causes, come what may.* Bishop categorically excludes agent causation, or causation by mind, from all biological explanation, and restricts the inference of intelligent agency to human activities. As he argues, "an intelligent agent is a presupposition external to cellular and evolutionary biology; intelligence has to be brought in from the outside."[12] Thus, if ID proposes agent causation to explain any biological event, it violates the well-defined boundaries of natural science—a violation, Bishop asserts, that "biologists rightly object to." The rule of MN has been broken.

Notice, first, that Bishop completely misunderstands the basis of Meyer's case for intelligent design. True, the intelligent agency that Meyer invokes to explain the origin of the information present in animal forms is "external" to the present operation of cells in those animals, just as the intelligence responsible for the design of a laptop computer is external to it. But that does not mean that Meyer "presupposes" that an agent "external to cellular and evolutionary biology" caused the origin of the information that arose in the Cambrian explosion of animal life. Instead, Meyer *infers* that a designing intelligence external to the features of cells and animals generated that information, and he does so based upon our knowledge of cause and effect and information-rich structures present in living systems. Since, as he argues, intelligence or mental activity is the only known cause of the origin of large amounts of functional or specified information, especially when that information is found in a digital form, the origin of the enormous amount of specified information that arose in the Cambrian period is best explained by the activity of a designing intelligence. Intelligence is not presupposed; it is inferred based upon what we know about the cause, indeed the only known cause, of specified information.

Notice too that Bishop's formulation of MN renders the evidence itself wholly irrelevant. If scientists must provide material or physical explanations for any phenomenon, whatever the evidence, then that is where they must remain, chained to the bench—even if the evidence strongly indicates design. This *a priori* formulation of MN makes a farce of empirical investigation, because the outcome of any research could never be in doubt: *some material or physical cause must be affirmed as the explanation.* If you don't find one, try harder; just keep looking until you do. That is what scientists have (allegedly) always done.

Bad History Makes for Bad Philosophy

But scientists haven't always done that, nor (as we'll explain below) do they follow MN today, except when keeping their boots firmly planted on the necks of ID proponents. As Bishop's own scholarly paper on MN shows[13]—see for instance his endnote 36, on Robert Boyle's view of the intelligent design of animals—leading figures in the Scientific Revolution did not see themselves as bound to strictly material or physical explanations. Isaac Newton, for example, made arguments for intelligent design in both the *Opticks* and the *Principia*. In the "General Scholium" to the *Principia*, he argued for the intelligent design of the solar system based upon the fine-tuning of the position of planets. As he stated:

> Though these bodies may indeed continue in their orbits by the mere laws of gravity, yet they could by no means have at first derived the regular position of the orbits themselves from those laws. Thus, this most beautiful system of the sun, planets, and comets, could only proceed from the council and dominion of an intelligent and powerful Being.[14]

Indeed, as an abundance of historical data affirms, MN became a putative philosophical convention of biology only after a protracted struggle. And one doesn't have to fight for a doctrine that everyone already accepts.

Moreover, it is impossible to make sense of the Darwinian Revolution if we assert that MN governed scientific explanation for centuries before Darwin's birth. In the "Introduction" to the *Origin of Spe-*

cies, Darwin surveys the landscape of existing scientific opinion—one can almost hear him drawing in his breath with apprehension—about "the view which most naturalists entertain, and which I formerly entertained—namely, each species has been independently created."[15] Looking around, Darwin would have needed no argument.

Were these "creationist" naturalists doing science? The answer is yes, unless one begs the question and identifies science with MN. These scientific contemporaries of Darwin, however, weren't conducting their investigations under the strictures of MN. In 1859, intelligent design was not only a live empirical possibility, but was generally *thought to be the best explanation of the origin of living forms*, thereby compelling Darwin to mount his "one long argument" against it. If MN already ruled, Darwin would have needed no argument.

The pre- (and post-) Darwinian existence of good science done without the strictures of MN shows that the rule is not necessary for the discovery or systematization of empirical knowledge. At bottom, the only real motivation for holding to MN is to keep the bad guys at bay, as an all-purpose defeater for ideas like intelligent design, especially when the data may not be cooperating. And all-purpose defeaters, while handy in many a difficult moment, eventually reveal themselves to be the cheats they are. Who wants to play in a soccer league where one team always wins, whatever the score on the field?

In short, MN never was the way science was always done. Science— empirical inquiry—pretty much takes care of itself, as long as curiosity, the evidence, and testability are given half a chance.

A Rule Honored in the Breach

Nor does MN govern today, except in official contexts (such as federal courts or statements from national science organizations) where definitions are required for demarcation purposes, to determine whether any idea passes muster as "science." Above, we noted that MN is a *putative* rule for biology—"putative" (that is, supposed but not actual) insofar as the content and practice of the science exhibit the widespread use of theological concepts and categories.

It is a little-remarked but nonetheless deeply significant irony that evolutionary biology is the most theologically entangled science going. Open a book like Jerry Coyne's *Why Evolution Is True*[16] or John Avise's *Inside the Human Genome*,[17] and the theology leaps off the page. A wise creator, say Coyne, Avise, and many other evolutionary biologists, would not have made this or that structure; therefore, the structure evolved by undirected processes. Coyne and Avise, like many other evolutionary theorists going back to Darwin himself, make numerous "God-wouldn't-have-done-it-that-way" arguments, thus predicating their arguments for the creative power of natural selection and random mutation on implicit theological assumptions about the character of God and what such an agent (if He existed) would or would not be likely to do.

Now, the usual response to criticism of this type of theologically grounded argument claims that evolutionary biology has been forced into its extra-scientific entanglements by stubborn religious opposition to the theory of evolution. The creationists started the fight, this view holds, so it's not surprising that evolutionary biologists need to push back using the terms and categories of creationists themselves.

Authors such as Coyne or Avise, however, hold that the apparently imperfect or suboptimal features of organisms provide *objective evidence* for undirected evolution. Presumably the standing of these features as *evidence* for evolution is not conditional on the presence, somewhere in the room, of a creationist or two saying otherwise. Put another way, Coyne and Avise would offer the same features as evidence to a science seminar populated by intelligent beings (aliens, let's say) with no concept of God or theology. Scientific evidence doesn't change its epistemic complexion depending on the audience or rhetorical context at hand.

If so—and Coyne has consistently defended the theological propositions in his book as fully empirical—then the very content of evolutionary theory rests on theological assumptions, borrowed or not. Philosopher of science Steven Dilley has carefully analyzed this situation[18] with respect to one of the most famous texts in twentieth-century biology, Theodosius Dobzhansky's essay "Nothing in biology makes sense except in the light of evolution."[19]

Although its title is widely cited as an aphorism, the text of Dobzhansky's essay is rarely read. It is, in fact, a theological treatise. As Dilley observes:

> Strikingly, all seven of Dobzhansky's arguments hinge upon claims about God's nature, actions, purposes, or duties. In fact, without God-talk, the geneticist's arguments for evolution are logically invalid. In short, theology is essential to Dobzhansky's arguments.[20]

Eventually evolutionary biologists themselves will grasp this reality, with inescapable consequences for the validity of MN. If Dobzhansky's essay genuinely belongs to the explanatory patrimony of evolutionary biology, MN is not only descriptively false (as history), but proscriptively unsound—we shouldn't follow the rule even if we could. MN is a bad philosophy of science on all counts.

In *Signature in the Cell* and *Darwin's Doubt*, Steve Meyer himself provides an exhaustive refutation of those who would enshrine MN as a normative convention for science. He shows that attempts to justify MN using various demarcation criteria—such as observability, replicability, and testability—have failed. He also shows that, in any case, the theory of intelligent design is testable in at least three interrelated ways.

First, like other scientific theories concerned with explaining events in the remote past, intelligent design is testable by comparing its explanatory power with that of competing theories.

Second, Meyer shows that ID, like other historical scientific theories, is tested against our knowledge of the cause-and-effect structure of the world. Following Darwin himself and the geologist Charles Lyell, Meyer shows that scientific historical theories provide adequate explanations when they cite causes that are known to produce the effects in question. These:

> considerations of causal adequacy provide an experience-based criterion by which to test—accept, reject, or prefer—competing

historical scientific theories. When such theories cite causes that are known to produce the effect in question, they meet the test of causal adequacy; when they fail to cite such causes, they fail to meet this test.[21]

Third, he shows that intelligent design makes a number of specific predictions that differ from predictions made by the materialistic theories of evolution against which ID competes (see his Appendix A in *Signature in the Cell* for a discussion of ten such predictions). These predictions not only provide another way to test the theory of intelligent design; they have, in several striking cases, already served "to confirm the design hypothesis rather than its competitors."[22]

For readers unfamiliar with Meyer's critique of the use of MN in science, I recommend Chapters 18 and 19 in *Signature in the Cell* and Chapter 19 in *Darwin's Doubt*, where he provides a thorough refutation of the case for accepting methodological naturalism as a normative rule for science. Indeed, in asserting MN as normative for science, Bishop really doesn't engage Meyer's earlier refutation of the need for MN, let alone refute Meyer's arguments against allowing MN to stand as a rule of method.

A Philosophy of Science No One Needs

MN DOES nothing for science that science cannot do for itself. Seen in the bright light of day, MN turns out to be little more than an all-purpose defeater for unwelcome ideas—another "Press Button in Case of Emergency" doctrine of the sort that brings disrepute on the philosophy of science. If ID is untestable or empirically empty, as its critics claim, we won't need MN to establish that. ID will fail on its own terms.

If ID *is* testable, however, as Meyer convincingly argues, then MN can only be a philosophical obstacle shoved in the way of the empirical possibility of design, for reasons having nothing to do with open-ended scientific inquiry. Either way, MN is a pointless rule.

Science will be better off without a rule that no one needs, that few actually obey, and that limits the freedom of scientists to follow the evidence wherever it leads. On this final point, let's give Meyer himself the last word:

> [A]llowing methodological naturalism to function as an absolute "ground rule" of method for all of science would have a deleterious effect on the practice of certain scientific disciplines, especially the historical sciences. In origin-of-life research, for example, methodological naturalism artificially restricts inquiry and prevents scientists from exploring and examining some hypotheses that might provide the most likely, best, or causally adequate explanations. To be a truth-seeking endeavor, the question that origin-of-life research must address is not, "Which materialistic scenario seems most adequate?" but rather, "What actually caused life to arise on earth?" Clearly, one possible answer to that latter question is this: "Life was designed by an intelligent agent that existed before the advent of humans." If one accepts methodological naturalism as normative, however, scientists may never consider this possibly true hypothesis. Such an exclusionary logic diminishes the significance of any claim of theoretical superiority for any remaining hypothesis and raises the possibility that the best "scientific" explanation (according to methodological naturalism) may not be the best in fact.[23]

Notes

1. Robert Bishop, "Meyer's Inference to Intelligent Design as the Best Explanation (Reviewing *Darwin's Doubt*: Robert Bishop, Part 3)," *The BioLogos Forum*, September 8, 2014), http://biologos.org/blog/meyers-inference-to-intelligent-design-as-the-best-explanation-reviewing-da.

2. Robert Bishop, "Final Assessments (Reviewing *Darwin's Doubt*: Robert Bishop, Part 4)," *The BioLogos Forum*, September 9, 2014, http://biologos.org/blog/final-assessments-reviewing-darwins-doubt-robert-bishop-part-4.

3. Hans Halvorson, "Why methodological naturalism?," *PhilSci-Archive* (September 2, 2014), http://philsci-archive.pitt.edu/11003/.

4. Maarten Boudry et al., "How Not to Attack Intelligent Design Creationism: Philosophical Misconceptions About Methodological Naturalism," *Foundations of Science* 15 (2010): 227–244; Maarten Boudry et al., "Grist to the Mill of Anti-evolutionism: The Failed Strategy of Ruling the Supernatural Out of Science by Philosophical Fiat," *Science & Education* 21 (2012): 1151–65.

5. Sahotra Sarkar, "The science question in intelligent design," *Synthese* 178 (2011): 291–305.

6. Bradley Monton, *Seeking God in Science: An Atheist Defends Intelligent Design* (Peterborough, ON: Broadview Press, 2009).

7. *Darwin's Doubt*, 19.

8. Ibid., 385.

9. Bishop, "Meyer's Inference to Intelligent Design," Part 3.

10. Ibid.

11. Ibid.

12. Ibid.

13. Robert C. Bishop, "God and Methodological Naturalism in the Scientific Revolution and Beyond," *Perspectives on Science and Christian Faith* 65:1 (2013): 10–23, http://www.asa3.org/ASA/PSCF/2013/PSCF3-13Bishop.pdf.

14. Isaac Newton, General Scholium to the *Principia* (1687), http://isaac-newton.org/general-scholium/.

15. Charles Darwin, *The Origin of Species* (1859 ed.), 6. At Darwin Online, http://darwin-online.org.uk/Variorum/1859/1859-6-dns.html.

16. Jerry Coyne, *Why Evolution Is True* (New York: Viking, 2009).

17. John Avise, *Inside the Human Genome: A Case for Non-Intelligent Design* (Oxford: Oxford University Press, 2010).

18. Steve Dilley, "Nothing in biology makes sense except in the light of theology?," *Studies in History and Philosophy of Biological and Biomedical Sciences* 44 (2013): 774–86, http://www.ncbi.nlm.nih.gov/pubmed/23890740.

19. Theodosius Dobzhansky, "Nothing in biology makes sense except in the light of evolution," *American Biology Teacher* (March 1973): 125–29.

20. Dilley, "Nothing in Biology," 774.

21. Stephen C. Meyer, *Signature in the Cell: DNA and the Evidence for Intelligent Design* (New York: HarperOne, 2009), 405.

22. Ibid.

23. Ibid., 437.

38.

MISTAKING INTELLIGENT DESIGN FOR A GOD-OF-THE-GAPS ARGUMENT

Casey Luskin

BIOLOGOS'S CONCLUDING ARTICLE IN ITS SERIES RESPONDING TO *Darwin's Doubt* was by theologian and philosopher Alister McGrath, the Andreas Idreos Professor at Harris Manchester College, University of Oxford. I'm a fan of McGrath's writings, but when it comes to intelligent design, there are problems. He has made a long series of inaccurate charges that ID is a "God of the gaps" argument.[1] Oddly, McGrath's piece for BioLogos isn't about *Darwin's Doubt* at all. In fact, it doesn't mention intelligent design. Rather, it's a transcript from a talk he gave, to which someone added the title, "Big Picture or Big Gaps? Why Natural Theology Is Better than Intelligent Design." However BioLogos evidently intends McGrath's piece as a response to intelligent design, so I'll treat the criticisms there accordingly. McGrath frames his critique as follows:

> My own approach is not to retreat into explanatory gaps. There are those who say (and perhaps I caricature or mis-say what they say), "Well, you know, science can't explain that. But if there were

a god, he could. Therefore, what science can't explain—that is a good reason for believing in God." And part of me wants to say, "Yes!" to that. But part of me also wants to say, well, this is not a very good idea, and leaves us bereft of the richness of a vision of God. It kind of implies that you believe in God because of the tiny little holes in somebody else's explanation, which you think you can explain better in brackets—at least for the time being. For me, it's not about saying, "Oh look! There's a gap there, and that's where God comes in!" No, no, no, it's about the big picture. That is what makes us think that the Christian faith makes sense of things.[2]

Before going further, it's helpful to understand what an early "coiner" of the phrase "God of the gaps," Dietrich Bonhoeffer, wrote in defining the concept:

> [H]ow wrong it is to use God as a stop-gap for the incompleteness of our knowledge. If in fact the frontiers of knowledge are being pushed further and further back (and that is bound to be the case), then God is being pushed back with them, and is therefore continually in retreat. We are to find God in what we know, not in what we don't know; God wants us to realize his presence, not in unsolved problems but in those that are solved.[3]

By that measure, intelligent design is *not* in fact a "God of the gaps" argument! I first encountered this comment from Bonhoeffer in Douglas Ell's book *Counting to God*, which rightly praised Bonhoeffer's reasonable argument.[4] ID does not find God, or evidence of design by any intelligent being, in "what we don't know" but rather in "what we know." The design inference is fundamentally grounded in our experience-based observations that high levels of complex and specified information (CSI) come only from intelligence. (For simplicity's sake, I will refer to high levels of CSI as "information," though I recognize that there are other types of "information.") We find evidence for design *in what we know* about the causes of new information. Intelligent design is an answer to the question of the origin of information.

Well, what if ID were a "gaps"-based argument? Such an argument would say, "Natural selection and random mutation cannot produce new information; therefore intelligent design is correct."

But there's a big and crucial difference between that argument and the actual case for intelligent design made by ID proponents. A genuine argument for intelligent design says something like, "Natural selection and random mutation cannot produce new information. Intelligent agency, uniquely in our experience, can produce new information. Therefore intelligent design is the better explanation for the information we see in life." This is not a gaps-based argument. It's a positive argument, based upon finding in nature the type of information that in our experience only comes from intelligence. Stephen Meyer frames the basic logic this way in *Darwin's Doubt*:

> Major Premise: If intelligent design played a role in the Cambrian explosion, then feature (X) known to be produced by intelligent activity would be expected as a matter of course.
>
> Minor Premise: Feature (X) is observed in the Cambrian explosion of animal life.
>
> Conclusion: Hence, there is reason to suspect that an intelligent cause played a role in the Cambrian explosion.[5]

You'd never find that kind of logic in a "gaps-based" argument. And as we'll see, there's a lot more positive content to the design inference.

McGrath's Odd Formulation of a "Gaps-Based" Argument

In the classical criticism of a "gaps-based" argument, "god" has no positive explanatory value whatsoever, other than to fill in and make up for the failure of some scientific explanation. But McGrath is attacking a slightly different formulation, which, in his description, says "science can't explain that. But if there were a god, he could."[6] In saying that a "god... could" explain something, McGrath apparently tries to add a positive component. Now, McGrath is no longer attacking a strictly negative "gaps-based" argument. Instead he is attacking an argument with some (though not much) positive explanatory value. Is McGrath necessarily justified in rejecting this sort of "gaps-based" argument?

No, not necessarily. Given that BioLogos frames McGrath's article as a broadside against Meyer's arguments for intelligent design, let's say that he intends to reject the following argument: "Science can't explain that. But intelligent agency could. Therefore, what science can't explain—that is a good reason for believing in intelligent design." Such an argument, which I am not making, would dramatically understate the explanatory power of intelligent design. But insofar as it does resemble the argument for ID, McGrath is unjustified in rejecting it. Why? Because the added word "could" means that ID has some level of positive explanatory value, meaning it's not simply a "gaps-based" argument. But just how much positive explanatory value does intelligent design offer? A huge amount.

McGrath's formulation dramatically understates the explanatory power of intelligent design. In *Darwin's Doubt*, Stephen Meyer doesn't just say that intelligent design "could" explain the information in life; he describes many specific properties of life and the Cambrian explosion that require an explanation that could only be intelligent design. Meyer looks at both the nature of the Cambrian animals themselves, and their appearance in the fossil record, and finds a whole array of specific features that are only explained by intelligent design. Meyer finds that "the cause of the origin of the new animal forms in the Cambrian explosion must be capable of" the following:

- Generating new form rapidly
- Generating a top-down pattern of appearance
- Constructing, not merely modifying, complex integrated circuits.

He goes on to say that "any explanation for the origin of the Cambrian animals must identify a cause capable of generating":

- Digital information
- Structural (epigenetic) information
- Functionally integrated and hierarchically organized layers of information.[7]

Let's look at some quotations to see how Meyer provides positive evidence to show that intelligent agents produce those features:

Generating New Form Rapidly:

Intelligent agents have foresight. Such agents can determine or select functional goals before they are physically instantiated. They can devise or select material means to accomplish those ends from among an array of possibilities. They can then actualize those goals in accord with a preconceived design plan or set of functional requirements. Rational agents can constrain combinatorial space with distant information-rich outcomes in mind.[8]

And elsewhere:

Intelligent agents sometimes produce material entities through a series of gradual modifications (as when a sculptor shapes a sculpture over time). Nevertheless, intelligent agents also have the capacity to introduce complex technological systems into the world fully formed. Often such systems bear no resemblance to earlier technological systems—their invention occurs without a material connection to earlier, more rudimentary technologies. When the radio was first invented, it was unlike anything that had come before, even other forms of communication technology. For this reason, although intelligent agents need not generate novel structures abruptly, they can do so. Thus, invoking the activity of a mind provides a causally adequate explanation for the pattern of abrupt appearance in the Cambrian fossil record.[9]

Generating a Top-down Pattern of Appearance:

"Top-down" causation begins with a basic architecture, blueprint, or plan and then proceeds to assemble parts in accord with it. The blueprint stands causally prior to the assembly and arrangement of the parts. But where could such a blueprint come from? One possibility involves a mental mode of causation. Intelligent agents often conceive of plans prior to their material instantiation—that is, the preconceived design of a blueprint often precedes the assembly of parts in accord with it. An observer touring the parts section of a General Motors plant will see no direct evidence of a prior blueprint for GM's new models, but will perceive the basic design plan immediately upon observing the finished product at the end of the assembly line. Designed systems, whether automobiles, airplanes, or computers, invariably manifest a design plan that preceded their first material instantiation. But the parts do not generate the whole. Rather, an idea of the whole directed the assembly of the parts.[10]

Constructing, Not Merely Modifying, Complex Integrated Circuits:

Integrated circuits in electronics are systems of individually functional components such as transistors, resistors, and capacitors that are connected together to perform an overarching function...

[I]n our experience, complex integrated circuits—and the functional integration of parts in complex systems generally—are known to be produced by intelligent agents—specifically, by engineers. Moreover, intelligence is the only known cause of such effects. Since developing animals employ a form of integrated circuitry, and certainly one manifesting a tightly and functionally integrated system of parts and subsystems, and since intelligence is the only known cause of these features, the necessary presence of these features in developing Cambrian animals would seem to indicate that intelligent agency played a role in their origin.[11]

Generating New Digital Information:

Intelligent agents, due to their rationality and consciousness, have demonstrated the power to produce specified or functional information in the form of linear sequence-specific arrangements of characters. Digital and alphabetic forms of information routinely arise from intelligent agents. A computer user who traces the information on a screen back to its source invariably comes to a mind—a software engineer or programmer. The information in a book or inscription ultimately derives from a writer or scribe. Our experience-based knowledge of information flow confirms that systems with large amounts of specified or functional information invariably originate from an intelligent source. The generation of functional information is "habitually associated with conscious activity." Our uniform experience confirms this obvious truth.[12]

And elsewhere:

Rational agents can arrange both matter and symbols with distant goals in mind. They also routinely solve problems of combinatorial inflation. In using language, the human mind routinely "finds" or generates highly improbable linguistic sequences to convey an intended or preconceived idea. In the process of thought, functional objectives precede and constrain the selection of words, sounds, and symbols to generate functional (and meaningful) sequences from a vast ensemble of meaningless alternative possible combinations of sound or symbol. Similarly, the construction of complex technological objects and products, such as bridges, circuit boards, engines, and software, results from the application of goal-directed constraints. Indeed, in all functionally integrated complex systems where the cause is known by experience or observation, designing engineers or other intelligent agents applied constraints on the possible arrangements of matter to limit possibilities in order to produce improbable forms, sequences, or structures. Rational agents have repeatedly demonstrated the capacity to constrain possible outcomes to actualize improbable but initially unrealized future functions. Repeated experience affirms that intelligent agents (minds) uniquely possess such causal powers.[13]

Generating New Structural (Epigenetic) Information and Constructing Functionally Integrated and Hierarchically Organized Layers of Information:

After noting that "the role of epigenetic information provides just one of many examples of the hierarchical arrangement (or layering) of information-rich structures, systems, and molecules within animals," Meyer writes:

> The highly specified, tightly integrated, hierarchical arrangements of molecular components and systems within animal body plans also suggest intelligent design. This is, again, because of our experience with the features and systems that intelligent agents—and only intelligent agents—produce. Indeed, based on our experience, we know that intelligent human agents have the capacity to generate complex and functionally specified arrangements of matter—that is, to generate specified complexity or specified information. Further, human agents often design information-rich hierarchies, in which both individual modules and the arrangement of those modules exhibit complexity and specificity—specified information as defined in Chapter 8. In-

dividual transistors, resistors, and capacitors in an integrated circuit exhibit considerable complexity and specificity of design. Yet at a higher level of organization, the specific arrangement and connection of these components within an integrated circuit requires additional information and reflects further design.

Conscious and rational agents have, as part of their powers of purposive intelligence, the capacity to design information-rich parts and to organize those parts into functional information-rich systems and hierarchies.[14]

Meyer concludes that "both the Cambrian animal forms themselves and their pattern of appearance in the fossil record exhibit precisely those features that we should expect to see if an intelligent cause had acted to produce them."[15] He sums his positive argument as follows:

> When we encounter objects that manifest any of the key features present in the Cambrian animals, or events that exhibit the patterns present in the Cambrian fossil record, and we know how these features and patterns arose, invariably we find that intelligent design played a causal role in their origin. Thus, when we encounter these same features in the Cambrian event, we may infer—based upon established cause-and-effect relationships and uniformitarian principles—that the same kind of cause operated in the history of life. In other words, intelligent design constitutes the best, most causally adequate explanation for the origin of information and circuitry necessary to build the Cambrian animals. It also provides the best explanation for the top-down, explosive, and discontinuous pattern of appearance of the Cambrian animals in the fossil record.[16]

Thus we see that Meyer identifies a breadth of features in both biology and the fossil record that are positively and uniquely explained by intelligence. There are specific positive reasons for inferring design, based upon our observations of intelligent agents and their products. We use those observations to generate expectations and predictions about what we should find if an intelligent agent was at work in generating the natural world. When we find those features in the natural world, and conclude that no other natural cause can explain them, we justifiably infer design. This makes ID the opposite of a "gaps-based" argument. The argument has multiple strong positive components, and without them, we cannot infer design.

Materialism of the Gaps

So why, despite this, does McGrath dismiss ID? Because his unwavering default position is to look exclusively to unguided material causes. He assumes methodological naturalism, and privileges material explanations in all circumstances regardless of their explanatory power. In McGrath's view, even if intelligent design has vast explanatory power, we should still not infer it, because we're filling a "gap" that ought to be filled by material causes.

In subjecting the scientific enterprise to the requirements of methodological naturalism, McGrath would force ID to operate under the presumption that natural causes always take precedence, regardless of whether they otherwise seem to fail. This itself is a "gaps-based" argument. McGrath assumes that material causes will eventually fill all the relevant gaps. This is not a real search for the best explanation. It's a search for the best explanation, provided that the explanation is naturalistic. It's materialism of the gaps.

Notes

1. Casey Luskin, "*The Dawkins Delusion*: Right on Dawkins, Wrong on Intelligent Design," *Evolution News & Views*, July 6, 2007, http://www.evolutionnews. org/2007/07/the_dawkins_delusion_right_on003789.html.

2. Alister McGrath, "Big Picture or Big Gaps? Why Natural Theology is better than Intelligent Design," *The BioLogos Forum*, September 15, 2014, http://biologos.org/ blog/big-picture-or-big-gaps-why-natural-theology-is-better-than-intelligent-des.

3. Dietrich Bonhoeffer, May 30, 1944, *Letters and Papers from Prison*, edited by Eberhard Bethge, translated by Reginald H. Fuller (New York: Touchstone, 1997), 331.

4. Casey Luskin, "*Counting to God*: New Book by Douglas Ell Introduces the Evidence for Intelligent Design," *Evolution News & Views*, June 10, 2014, http://www. evolutionnews.org/2014/06/counting_to_god086541.html.

5. Stephen C. Meyer, *Darwin's Doubt: The Explosive Origin of Animal Life and the Case for Intelligent Design* (New York: HarperOne, 2013), 351.

6. McGrath, "Big Picture."

7. *Darwin's Doubt*, 357–58.

8. Ibid., 362–63.

9. Ibid., 373, 375.

10. Ibid., 371–372.

11. Ibid., 364.

12. Ibid., 360.

13. Ibid., 362.

14. Ibid., 366.

15. Ibid., 379.

16. Ibid., 381.

39.

CLARIFYING ISSUES:

MY RESPONSE TO BIOLOGOS

Stephen C. Meyer

I APPRECIATE THE CLOSE READING AND CAREFUL EVALUATION OF MY book *Darwin's Doubt* by the authors of the multi-part review series published on the BioLogos website. I would like to thank the main reviewers of the book (Ralph Stearley, Robert Bishop, and Darrel Falk) for taking the time to read and review the book as well as BioLogos and its president Deborah Haarsma for their decision to highlight these reviews and their generous invitation to me to submit a response.[1] Anyone whose work receives such scrutiny, with such a breadth of coverage, will learn something, and I certainly have.

I have especially appreciated how the reviews in this recent series have unexpectedly clarified the nature of the disagreement between proponents of the theory of intelligent design and the proponents of theistic evolution (or evolutionary creation) associated with BioLogos. I—and many others—have long assumed that the debate between our two groups was mainly a scientific one about the adequacy of contemporary evolutionary theory. Surprisingly, the reviews collectively have shown that the main disagreement between ID proponents and BioLogos is not scientific, but philosophical and methodological.

They have revealed that the central issue dividing the BioLogos writers from intelligent design theorists concerns a principle known as methodological naturalism (MN). MN asserts that scientists must explain all events and phenomena by reference to strictly naturalistic or materialistic causes.

The principle forbids postulating the actions of personal agency, mind, or intelligent causation in scientific explanations and thus limits the explanatory toolkit of science. The principle of methodological naturalism is, of course, not a scientific theory nor an empirical finding, but an allegedly normative methodological rule, against which I have argued in depth, both in *Darwin's Doubt* (Chapter 19) and in my earlier book, *Signature in the Cell* (Chapters 18 and 19). My colleagues have also argued against MN in their responses to some of the BioLogos reviews of *Darwin's Doubt*. (See Chapters 36 and 37 in this book.)

Recall that *Darwin's Doubt* argues that intelligent design provides the best explanation for the origin of the genetic (and epigenetic) information necessary to produce the novel forms of animal life that arose in the Cambrian period. In making this case, I show first that neither the neo-Darwinian mechanism of natural selection acting on random mutations, nor more recently proposed mechanisms of evolutionary change (species selection, self-organization, neutral evolution, natural genetic evolution, etc.) are sufficient to generate the biological information that arises in the Cambrian period. (See *Darwin's Doubt*, Chapters 15 and 16.) Instead, I show—based upon our uniform and repeated experience—that only intelligent agents have demonstrated the power to generate the kind of functional information that is present in biological systems (and that arises with the Cambrian animals). Thus, I conclude that the action of a designing intelligence provides the best ("most causally adequate") explanation for the origin of that information.

Paleontologist Ralph Stearley and Geneticist Darrel Falk

Now, ONE might have expected that Ralph Stearley,[2] a paleontologist, and Darrel Falk[3], a geneticist, both of whom have extensive knowledge of evolutionary theory, would have critiqued the main scientific argument of *Darwin's Doubt* on scientific grounds. In particular, one might have expected them to argue that either the neo-Darwinian mechanism or some other evolutionary mechanism does have the creative power to produce the information necessary to build new forms of animal life. Instead, except for raising a few minor objections about incidental scientific matters, both acknowledged that evolutionary theory has left the problem of the Cambrian explosion unsolved—i.e., that the mutation/natural selection mechanism lacks the creative power to account for macroevolutionary innovations in the history of life.

Falk, for instance, wrote that *Darwin's Doubt* identifies "one of the great mysteries in evolutionary biology today," namely, the origin of animal form.[4] Falk observed that this problem has never really been addressed by neo-Darwinian theory, and reflected on his own experiences as a college teacher of evolution discovering the shortcomings of textbook theory when confronted with the origin of complex animal evolution. He added that the process of natural selection, important as it may be in certain contexts, is not the "driving mechanism" of macroevolutionary change, and thus that the mystery of the Cambrian explosion still awaits a solution.

Of course, Falk himself rejects my proposed solution and my positive argument for intelligent design as the best explanation for what I call the "Cambrian information explosion." He contends that any such inference to intelligent design is premature. Nevertheless, Falk doesn't really offer any evidence or *scientific* reason for rejecting the positive argument of *Darwin's Doubt*. Indeed, it would be difficult for him to deny that intelligent agents possess the causal power to produce functional information. Is it possible, then, that his reluctance to consider intelligent design as the best or "most causally adequate" explanation stems from a tacit commitment to methodological naturalism? If inferences to intelligent design are perceived as breaking the rules of science then, of course, they will always be seen as premature.

Stearley also found value in the book's scientific analysis, saying that it "makes an argument that folks should think hard about" and indeed that he "resonate[s] with some of Meyer's arguments."[5] He is unhappy with aspects of my book and thinks I should have talked more about the small shelly fossils in the early Cambrian, something my colleagues and I have addressed in Chapters 13 and 14. But Stearley agreed with my critique of the adequacy of current evolutionary mechanisms for the origin of animal form. Thus, Stearley notes that I "developed a case for the inadequacy" of standard approaches.[6] On scientific grounds, therefore, relatively little of note separates us. In fact, Stearley admitted that he was "inclined to see design in nature,"[7] but he too demurred from affirming the design hypothesis, offering hesitant uncertainty in response to my positive case. Could it be that in Stearley's reluctance, we may, again, be seeing a tacit commitment to methodological naturalism?

Philosopher of Science Robert Bishop

OF THE three reviewers, Wheaton College philosopher of science Robert Bishop was the least persuaded by my arguments—but, interestingly, he was also the most explicitly committed to the principle of methodological naturalism.[8] Indeed, he objected to the thesis of the book precisely because it openly rejects (and violates) the principle of methodological naturalism.

Consequently, his four-part critique, by far the longest in the BioLogos series, said very little about my scientific arguments. He did argue that I was wrong to claim that newer models of evolutionary theory represent significant deviations from neo-Darwinian orthodoxy. Yet, notably, biologist Darrel Falk's review affirmed my assessment of these newer theories over and against Bishop's.

In any case, Bishop focused his critique on what he called my "rhetorical strategies," giving particular attention to philosophical issues concerning the legitimacy of design inferences in biology. In Bishop's judgment, intelligent design flagrantly violates the rule of methodological naturalism—a rule that he regards as normative for the practice of all natural science because he believes (incorrectly, as it turns out) that "methodological naturalism is the way scientific investigation has been done since before the time of the Scientific Revolution."[9] Indeed, as my colleague Paul Nelson pointed out in his response to Bishop's critique, Bishop badly misreads the history of science. The design arguments developed by Isaac Newton—in the *Opticks* and the *Principia*, for instance—alone contradict Bishop's claims.

Even so, Bishop correctly notes that methodological naturalism does categorically exclude consideration of inferences to the activity of non-physical entities or causes (i.e., intelligent agents or minds) in evolutionary or historical biology. These fields simply do not allow reference to the activity of intelligent agents. Bishop appears to justify this prohibition by claiming that "an intelligent agent is a presupposition external to cellular and evolutionary biology; intelligence has to be brought in from the outside"—a move that, in his view, would transgress the boundaries of natural science and that "biologists rightly object to."[10]

Of course, asserting that methodological naturalism prohibits design inferences and then justifying that prohibition by arguing that inferring intelligent design would transgress the boundaries of science *as determined by methodological naturalism,* is to argue in a circle.

In any case, by focusing his critique on the allegedly normative status of methodological naturalism, and my repudiation of that methodological convention as a normative rule for science, Bishop did not focus his critique on the scientific claims or analysis of the book.

Disagreements over Methodological Naturalism

THUS, BOTH Bishop's review (which challenged the methodological approach, but not the scientific analysis, of the book), and Falk and Stearley's reviews (both of which conceded my main scientific critique of evolutionary theory) have helped to clarify the true nature of our disagreement. Since I look forward to further dialogue with our colleagues at BioLogos, I regard these reviews as a constructive first step to further discussion of the key issues that separate us.

As we continue to our discussion, I hope we can address the central issue about which we disagree. As noted, I have developed a detailed critique of methodological naturalism in my published work. I have shown, for example, that the demarcation criteria typically offered as justifications for methodological naturalism invariably fail to distinguish the scientific status of intelligent design and competing evolutionary theories. I have also argued that the principle of methodological naturalism restricts the intellectual freedom of scientists and compels them to select materialistic explanations, whatever the evidence may indicate. As such, the principle impedes the truth-seeking (as opposed to convention-following) function of science.

Given my own skepticism about methodological naturalism, I would very much like to know what Darrel Falk and Ralph Stearley think about the principle and its alleged status as rule governing scientific reasoning. Their reviews express hints that design inferences in historical biology might be acceptable to them—yet those same reviews reveal a deep ambivalence about challenging the naturalistic premises of current evolutionary theory, or more fundamentally, about challenging MN itself.

Unfortunately, methodological naturalism is a demanding doctrine. The rule does not say "try finding a materialistic cause but keep intelligent design in the mix of live possibilities, in light of what the evidence might show." Rather, MN tells you that you simply must posit a material or physical cause, whatever the evidence. One cannot discover evidence of the activity of a designing mind or intelligence at work in the history of life because the design hypothesis has been excluded from consideration, even before considering the evidence.

Having a philosophical rule dictate that one may not infer or posit certain types of causes, whatever the evidence, seems an exceedingly odd way for science to proceed. Scientists tend to be realists about the power of evidence, but skeptics about philosophical barriers—which, if it is anything, the rule of MN surely is. Thus placing the detection of intelligent design out of the reach of scientific investigation looks like rigging a game before any players have taken the field.

In the debate about intelligent design, MN has compelled many scientists to dismiss evidence for intelligent agency as an explanation for phenomena, such as increases in functional digital information, that are known to be produced by one—and only one—kind of cause, namely, intelligent activity. Proponents of intelligent design reject this restriction precisely because it compromises the truth-seeking function of science. We insist that scientists should seek the best explanation, based upon our knowledge of the evidence and the causal powers of competing explanatory entities, not seek the best explanation only among an artificially restricted set of options. Our BioLogos colleagues appear to disagree.

This issue won't be easy to resolve, because, while in scientific disputes an impasse can be broken by new evidence, *MN keeps the evidence itself out of the discussion*. If ever a rule of method deserved to be tossed onto the rubbish heap of history, now is the time for MN to be sent in that direction. Natural science has nothing to fear from allowing scientists to consider evidence for design hypotheses because (given the general cultural climate) their rigorous testing is assured, as the vigorous attacks on notions such as "irreducible complexity" and "specified complexity" over the past two decades have already shown. Many scientists have attempted to burnish their scientific standing by publishing challenges to claims made by proponents of intelligent design.

In a similar vein, I invite our colleagues at BioLogos to engage and reply to our critique of the principle of methodological naturalism—to defend, rather than just assert (as even Bishop mainly did), the normative status of MN. Offering such a defense will doubtless afford further opportunities for clarification and discussion of the key issues.

Constructive dialogue between parties with significant disagreements can, in the best case, expose both common ground and the true nature of those disagreements. The reviews published by BioLogos have done both—a fact for which I, as the author of the book under discussion, am genuinely grateful.

Notes

1. Stephen C. Meyer, "Clarifying Issues: My Response to the BioLogos Series reviewing 'Darwin's Doubt,'" *The BioLogos Forum*, January 19, 2015, http://biologos.org/blog/clarifying-issues-my-response-to-the-biologos-series-reviewing-darwins-doub.

2. Ralph Stearley, "Reviewing *Darwin's Doubt*," *The BioLogos Forum*, August 26, 2014, http://biologos.org/blog/reviewing-darwins-doubt-ralph-stearley; Ralph Stearley, "The Cambrian Explosion: How Much Bang for the Buck?," *Perspectives on Science and Christian Faith* 65 no. 4 (2013): 245–257, http://www.asa3.org/ASA/PSCF/2013/PSCF12-13Stearley.pdf.

3. Darrel Falk, "Thoughts on *Darwin's Doubt* (Reviewing *Darwin's Doubt*: Darrel Falk, Part 1)," *The BioLogos Forum* September 9, 2014, http://biologos.org/blog/thoughts-on-darwins-doubt-reviewing-darwins-doubt-darrel-falk-part-1; Darrel Falk, "Further Thoughts on *Darwin's Doubt* after Reading Bishop's Review (Reviewing *Darwin's Doubt*: Darrel Falk, Part 2)," *The BioLogos Forum*, September 11, 2014, http://biologos.org/blog/thoughts-on-darwins-doubt-reviewing-darwins-doubt-darrel-falk-part-2.

4. Falk, "Thoughts on *Darwin's Doubt*."

5. Stearley, "Cambrian Explosion."

6. Ibid.

7. Ibid.

8. Robert C. Bishop, "The Extended Synthesis (Reviewing *Darwin's Doubt*: Robert Bishop, Part 1)," *The BioLogos Forum*, September 1, 2014, http://biologos.org/blog/the-grand-synthesis-reviewing-darwins-doubt-robert-bishop-part-1; Robert C. Bishop, "The Extended Synthesis (Reviewing *Darwin's Doubt*: Robert Bishop, Part 2)," *The BioLogos Forum*, September 2, 2014, http://biologos.org/blog/two-rhetorical-strategies-reviewing-darwins-doubt-robert-bishop-part-2; Robert Bishop, "Meyer's Inference to Intelligent Design as the Best Explanation (Reviewing *Darwin's Doubt*: Robert Bishop, Part 3)," *The BioLogos Forum*, September 8, 2014, http://biologos.org/blog/meyers-inference-to-intelligent-design-as-the-best-explanation-reviewing-da; Robert Bishop, "Final Assessments (Reviewing *Darwin's Doubt*: Robert Bishop, Part 4)," *The BioLogos Forum*, September 9, 2014, http://biologos.org/blog/final-assessments-reviewing-darwins-doubt-robert-bishop-part-4.

9. Bishop, "Meyer's Inference."

10. Ibid.

40.

WALKING IT BACK?

Stephen C. Meyer

IN HER CONCLUSION TO BIOLOGOS'S TEN-PART REVIEW SERIES OF MY
book, *Darwin's Doubt*, BioLogos President Deborah Haarsma suggested that I mischaracterized the perspective of the organization's
reviewers in my response to them.[1] She asserts that, contrary to my
portrayal, the BioLogos scientists who reviewed *Darwin's Doubt* do not
regard the Cambrian explosion as an unsolved problem from the standpoint of evolutionary theory.

After re-reading what the BioLogos reviewers actually wrote, I
stand by my original assessment. As the record shows, the BioLogos
scientists reviewing my book not only acknowledged the inadequacy of
the neo-Darwinian mechanism as an explanation for the origin of the
animal body plans that arose in the Cambrian period, but they also acknowledged that no other known evolutionary mechanism can explain
this event.

Recall that my main argument in *Darwin's Doubt* is that the origin of
the genetic (and epigenetic) information necessary to produce the novel
forms of animal life that arose in the Cambrian period is best explained
by intelligent design. To make this case, I showed first that neither the
neo-Darwinian mechanism of natural selection acting on random muta-

tions, nor more recently proposed mechanisms of evolutionary change (such as self-organization, neutral evolution, natural genetic engineering, etc.) are sufficient to generate the biological information that arises in the Cambrian explosion. (See *Darwin's Doubt*, Chapters 15–16.) Instead, I show—based upon our uniform and repeated experience—that only intelligent agents have demonstrated the power to generate the functional information of the kind that is present in biological systems (and that arises with the Cambrian animals). Thus, I conclude that the action of a designing intelligence provides the best ("most causally adequate") explanation for the origin of that information.

In my response to the BioLogos critical review series, I noted that the main scientific reviewers (Falk and Stearley) had actually agreed with my scientific assessment of the inadequacy of the neo-Darwinian mechanism and that they had also acknowledged that no other known evolutionary mechanism can (yet, at least) account for the origin of novel animal body plans that arose in the Cambrian. Thus, I suggested that the series had unexpectedly clarified the true nature of the disagreement between proponents of intelligent design and the BioLogos scientists (though philosopher Robert Bishop is a partly different matter, see below). In particular, I suggested that our disagreement derives less from differing assessments of the current status of evolutionary theory (i.e., the science) than from differing views about the *rules of science* and, specifically, whether those rules preclude consideration of the design hypothesis and require scientists to search into the indefinite future for some materialistic cause or process as the best explanation for all phenomena and events, whatever the evidence. In other words, I suggested that the series had clarified that our real disagreement mainly concerns the legitimacy of design inferences and the closely related issue of whether methodological naturalism should be regarded as a normative convention governing all scientific theorizing.

In her reply to my response, Haarsma maintained that I had mischaracterized the views of Falk and Stearley. Haarsma noted that Darrel Falk and Ralph Stearley did acknowledge the inadequacy of the standard neo-Darwinian mechanism of natural selection and random mutation as an explanation for the origin of novel forms of animal life. But she also explicitly stated that the BioLogos reviewers, including the two scientist reviewers, deny that the Cambrian explosion is an unsolved problem in evolutionary theory. As she wrote, "While the authors agree with Meyer and mainstream biologists that one mechanism of evolution (natural selection) is insufficient by itself to explain the development of animal body plans, they did not call the Cambrian explosion 'unsolved' or 'awaiting a solution.'"[2] Instead, Haarsma suggested the opposite, namely, that both BioLogos scientists affirm the adequacy of *other* evolutionary mechanisms, perhaps in conjunction with the mutation/selection mechanism, as an explanation for the origin of the animal body plans that arise in the Cambrian period—i.e., that the so-called "extended synthesis" has solved that problem.

Yet neither of the BioLogos scientists actually wrote this. Indeed, Darrel Falk—the only biologist among the team of reviewers—clearly said the opposite and Ralph Stearley characterized the current status of evolutionary theory in much the same way as did Falk.

Falk on the Mystery of the Cambrian Explosion

Much as I do in *Darwin's Doubt*, Darrel Falk calls the Cambrian explosion a "mystery" (actually a "big mystery") and acknowledges that none of the recently proposed evolutionary mechanisms or models has provided an adequate account of the origin of novel animal form. Referring specifically to these recently proposed (i.e., post-neo-Darwinian) mechanisms, he wrote: "Stephen is right, that none of the other models fit the bill in a fully satisfactory manner yet."[3]

In making this concession to the critical or negative part of my argument, Falk does add the important qualifying word "yet." But in saying that current evolutionary models have not yet solved the problem he is saying precisely what I said that he said and precisely what Haarsma denies that he said. Saying a problem has not yet been solved in a satisfactory way and saying that it "awaits its solution" are about as close to equivalent expressions as can be formulated in English.

Of course, Falk also expresses optimism about what he expects evolutionary theory to achieve in the future. As he states:

> As Douglas Erwin elegantly argues in his 2011 paper, there must have been something different taking place as the system was being put in place 550 million or so years ago. I think figuring that out will turn out to be one of the most fascinating pieces of puzzle-solving that molecular biology has ever done. However, unlike Stephen, not only do I think this research is not at a dead-end, I think it will turn out to be among the most exciting frontiers in biological research over the next couple of decades. The work, as most developmental biologists see it, has only just begun, and it is the kind of thing that happens at this cutting edge stage, which makes science so much fun. I'm with Ralph Stearley on this: to study the diversity of life and the mechanisms which characterize it is to be enraptured in joy.[4]

I admire Darrel's enthusiasm for scientific investigation and duly noted his confidence in what evolutionary biology will one day discover. In my response, I acknowledged that Darrel expects biologists to find a materialistic evolutionary process that can account for the origin of animal body plans—that was presupposed in my questions about the extent to which he is committed to methodological naturalism. Nevertheless, issuing promissory notes about the creative power of some yet undiscovered materialistic process is not the same thing as affirming that evolutionary biology has in fact discovered such a process with the capacity to generate novel animal body plans or that the mystery of the Cambrian explosion has been solved.

Stearley on Scientists "Looking to Build" an Adequate Model

Ralph Stearley similarly seems to believe that a "larger synthesis," which can explain the origin of animal life as "the outcome of biological processes," is in the works and forthcoming, but that biologists are still "looking to build" an adequate model, which is not yet sufficient or complete. Specifically, Stearley writes:

> while it is true that Goodwin and others believe that their discoveries pose a major challenge to neo-Darwinian orthodoxy, this does not cause them to abandon their *belief* that the history of life

can be explained as the outcome of biological processes! Indeed, many evolutionary biologists and paleontologists *are looking* to build the notions provided by morphogenetic fields and developmental constraints into a larger synthesis.[5]

Stearley characterizes evolutionary biologists as seeking to discover materialistic "biological processes" capable of accounting for the origin of morphological novelty, but he implies that the effort to produce this explanatory synthesis is a work in progress—that evolutionary biologists are still "looking to build" an adequate model. To the extent that he is clear in his own views, therefore, he does not present current evolutionary theory as having provided an adequate explanation for the origin of animal body plans or macro-evolutionary innovation generally. (The origin of these things is, of course, of significant concern to structuralists such as the late Brian Goodwin as well as representatives of the many post-neo-Darwinian schools of thought.)

That Stearley understands the Cambrian explosion to be an unsolved problem for evolutionary theory can be also clearly discerned in his parallel review of Douglas Erwin and James Valentine's 2013 book *The Cambrian Explosion* in the same essay in which his review of *Darwin's Doubt* appeared. Stearley quotes these two Cambrian experts approvingly as saying the Cambrian explosion represents "a tractable but unresolved problem."[6] He goes on to say:

> Erwin and Valentine admit that there is much yet to be deciphered concerning the Precambrian-Cambrian biotic transition. They see two major unresolved questions:
>
> > First, what evolutionary processes produced the gaps between the morphologies of the major clades? Second, why have the morphologic boundaries of these body plans remained relatively stable over the past half a billion years? (p. 330).[7]

In quoting Erwin and Valentine candidly acknowledging the lack of an adequate mechanism or known biological process (or processes) to account for "the Cambrian diversification event," Stearley does not dispute the judgment of these authorities, but instead reviews their book favorably. Thus, my representation of Stearley as viewing the Cambrian explosion as a problem "await[ing] a solution" seems to be entirely correct.

In any case, if Stearley (and the other BioLogos reviewers) do think that these problems have been solved, as Haarsma now suggests, they would in so claiming certainly contradict leading Cambrian experts like Erwin and Valentine. Earlier in the same book, where Erwin and Valentine note that major questions like "what evolutionary process produced the gaps between the morphologies of major clades?" are "unresolved,"[8] they also question whether "uniformitarian explanations can be applied to understand the Cambrian explosion."[9] Indeed, Erwin has pointedly taken a non-uniformitarian view of the evolution of body plans in which he maintains that no known biological process accounts for the Cambrian explosion of animal form—i.e., that *whatever* caused the Cambrian explosion is unlike any biological process observed today.[10]

If Haarsma is correct that Stearley and other BioLogos reviewers actually do think the Cambrian explosion has been adequately explained by some *known* evolutionary mechanism (or combination of mechanisms), then it would be reasonable to expect that the reviewers would have provided descriptions of how these alleged processes account for the origin of novel animal form in the Cambrian period. It would also be reasonable to expect that they would explain how such a known mechanism solves the specific problems discussed in *Darwin's Doubt*.

In particular, they might have offered an explanation for how some proposed mechanism (or combination of mechanisms) explains: (1) the origin of genetic information and novel proteins (given the size of the combinatorial sequence space that must be searched in the available evolutionary time); (2) the origin of epigenetic information (given that genetic mutations only act on genes, not epigenetic sources of information); (3) the origin of body plans (given that developmental mutations invariably produce embryonic lethals); and (4) the origin of novel gene regulatory networks (given that all known experimentally induced perturbations in such networks disrupt animal development). Neither Stearley nor any of the other BioLogos reviewers attempted to explain how known evolutionary mechanisms resolve these difficulties—a fact that reinforced my judgment that they were not claiming to have solved the problem of the origin of animal body plans in the Cambrian period, but viewed the problem as one "await[ing] a solution."[11]

Robert Bishop's Scientific Disagreement

In contrast to Falk and Stearley, philosopher Robert Bishop does seem to suggest that the problem of the origin of evolutionary novelty has been solved (or significantly minimized) by recognizing the role of new evolutionary mechanisms and by affirming that these additional mechanisms act in concert with the standard neo-Darwinian mechanism of natural selection and random mutation. Thus, he represents advocates of newer evolutionary theories and models as continuing to affirm the central importance of natural selection and random mutation, albeit in conjunction with other additional mechanisms.

For this reason, Haarsma is correct that Bishop (though not either Falk or Stearley) has a significant scientific disagreement with me about the current status and adequacy of evolutionary theory as an explanation. First, he presents newer evolutionary models as mere supplements to a basically sound core of neo-Darwinian theory—not, in many cases, radical departures from a failed theory, as I do. Second, he seems to affirm that these other evolutionary mechanisms acting in concert with the standard neo-Darwinian mechanism of mutation and natural selection are adequate to explain macro-evolutionary innovation.

Nevertheless, Darrel Falk, BioLogos's own biological expert, pointedly contradicted Bishop's judgment about the extent to which these models represent only modest extensions or supplements of neo-Darwinian theory. Instead, Falk confirmed my description of many of these models as representing radical departures from neo-Darwinism. And Falk is clearly correct about this. Indeed, many of these models repudiate crucial aspects of the neo-Darwinian synthesis, by denying, for example, the central importance of natural selection (as do neutral theorists), or the central role of random mutations (as do self-organizational theorists), or the random nature of mutations (as do advocates of natural genetic engineering).

Thus, advocates of these and other new models of evolutionary theory do not continue to think that natural selection and random mutation play a central role in evolutionary innovation, as Bishop seems to claim. Consequently, Bishop cannot be right that proponents of these new models see themselves as having solved the problem of the origin of evolutionary novelty by supplementing the selection and mutation mechanism with other mechanisms. Most proponents of these newer models see themselves as proposing new mechanisms to replace the mutation/selection mechanism as the key driver in evolutionary innovation. Since Falk criticized Bishop's depiction of the state of evolutionary theory (by pointing out that Bishop had incorrectly denied that these newer models represent radical departures from standard theory), I saw no need to belabor his criticism of Bishop's scientific judgment or to treat Bishop's view as characteristic of BioLogos's scientific position.

Why the Confidence in Materialistic Processes?

In any case, my interest in writing what I did was to probe exactly why Darrel and the other BioLogos reviewers appear so confident that materialistic processes *will* eventually prove sufficient to explain all phenomena in the history of life, including phenomena such as the origin of functional digital information that we know from experience to arise only from the activity of intelligent agents, and including events such as the Cambrian explosion that have long—since Darwin's time at least—resisted materialistic explanation.

In his review, Darrel himself explained the depth of the conceptual problems confronting evolutionary theory—whatever his other expressions of confidence in the eventual adequacy of a purely naturalistic approach to the problem.

For example, he did a nice job of explaining how the functionally integrated networks of genes and gene products that control key aspects of animal development resist perturbation, and thus why it is hard to envision one animal body plan arising from another, given what we know about the importance of these gene regulatory networks to animal development.

Given the depth of this and other related conceptual difficulties, such as the presence in developing animals of something akin to an integrated control system (or circuit), why not consider the possibility that systems at work in animal development bear witness to the designing agent that BioLogos scientists believe to be a reality? Despite his recognition of the depth of the conceptual problems confronting contemporary evolutionary theory, Darrel does not seem open to this possibility. My questions is simply: Why not?

I consider the recognition and detection of intelligent design to be a scientific possibility because of my study of the methodology of the historical sciences. The central methodological desideratum of the historical sciences—as pioneered in large part by Darwin himself—is the need to explain events in the remote past by reference to causes known from our present experience to have the power to produce the effects in question—i.e., Darwin's *vera causa* criterion or what Lyell called "causes now in operation."

Modern molecular biology has revealed that building animal body plans requires vast infusions of new functional genetic information (stored in a digital form). Modern developmental biology has shown the need for networks of genes and gene products that function as integrated control systems or circuits. Developmental biology has also revealed the importance of other sources of "epigenetic" information for building animal form and, consequently, the existence of a hierarchically organized information processing system at work in animal development.

Yet, we know of one, and only one, type of cause capable of producing these necessary conditions of building animal form. In our experience, digital information, integrated control systems (and circuitry) and hierarchically organized information processing systems invariably arise from intelligent causes—from conscious and rational activity. So why not consider the possibility that such a cause played a role in the origin of animal life?

The reason that Darrel and Ralph do not consider the possibility of design is not that they know of an evolutionary mechanism (or material process) that has demonstrated the power to produce these necessary features and conditions of building animal body plans. Nor can they point with any specificity to what a possible solution to the problem of the origin of animal form and biological information will look like in materialistic terms. Instead, at best, they can affirm, as Darrel does in his almost creedal statement at the end of his first review, that all phenomena can (or must) be explained by reference to natural laws. As he explains:

> I see no scientific, biblical, or theological reason to expect that [an intelligent agent might have acted discretely or discernibly in the history of life]. Natural processes are a manifestation of God's ongoing presence in the universe. The Intelligence in which I as a Christian believe, has been built into the system from the beginning, and it is realized through God's ongoing activity which is manifest through the natural laws. Those laws are a description of that which emerges, that which is a result of, God's ongoing presence and activity in the universe. I see no biblical, theological, or scientific reason to extend that to extra supernatural "boosts" along the way...[12]

Darrel's description of his philosophy and theology of nature is admirably clear. It amounts to the *a priori* conviction that during natural history God acts mainly (or exclusively) through secondary causes such that we are justified in seeking—into the indefinite future—only law-like material processes to explain natural phenomena, including the origin of fundamentally new forms of life and the origin of the information necessary to produce them. His philosophy of nature constitutes a tacit commitment to the idea that all phenomena and events in natural history can be (or should be) explained by reference to what theologians think of as "secondary causes." But that is just another way of expressing a commitment—perhaps a distinctively Christian way of expressing a commitment—to the principle of methodological naturalism. And that, of course, was exactly my point.[13]

Robert Bishop actually makes such a commitment explicit in his response to me, stating (incorrectly, as it turns out) that "methodological naturalism is the way scientific investigation has been done since before the time of the Scientific Revolution and is well-grounded theologically."[14] Given Falk's apparent commitment to some form of methodological naturalism (as shown above), Haarsma's argument that BioLogos as an organization "does not have a position on the use of the term" methodological naturalism would seem to be something of a red herring (and a bit disingenuous). It may be true that BioLogos has no official position on the use of the term, but the style of thinking (and limitation on scientific theorizing) that the term designates accurately describes the intellectual commitments of the BioLogos reviewers, and thus, may also help explain why those reviewers and ID proponents take different positions on the issue of design.

Of course, if scientists and scholars have effectively ruled out the possibility of the design hypothesis as part of science, then no amount of evidence will suffice to justify such a hypothesis (or inference) for those thus committed. Given that the BioLogos reviewers did not provide, or even point to, anything like a detailed alternative scientific explanation for the origin of novel animal body plans (and/or the information necessary to produce them), it seems clear that their reasons for affirming the eventual adequacy of some materialistic evolutionary processes have little to do with the current state of scientific evidence or theorizing. This suggests that their opposition to considering the design hypothesis may be based upon extra-evidential commitments about the desirability of explaining all phenomena by reference to purely materialistic or naturalistic processes—as the principle of methodological naturalism requires.

In any case, it is not at all clear that BioLogos has declined to take an official position on methodological naturalism. In their description of the theory of intelligent design on their website, BioLogos affirms its commitment to explaining all natural phenomena (including presumably the origin of life and novel forms of life) by reference to strictly natural causes. As the website explains:

> [Intelligent Design] claims that the existence of an intelligent cause of the universe and of the development of life is a testable scientific hypothesis. ID arguments often point to parts of scientific theories where there is no consensus and claim that the best solution is to appeal to the direct action of an intelligent designer. At BioLogos, we believe that our intelligent God designed the universe, but we do not see scientific or biblical reasons *to give up on pursuing natural explanations for how God governs natural phenomena.*[15]

Indeed, BioLogos writers have repeatedly affirmed the principle of methodological naturalism—as the preceding statement surely does—in numerous contexts.[16] Bishop critiqued my book precisely because it repudiates "methodological naturalism."

All this would seem to make it entirely fair to question the extent to which *a priori* commitments to this principle disincline the BioLogos reviewers from considering the evidence for, and the logical basis of, intelligent design as an explanation for various classes of evidence. By denying that these commitments, or at least intellectual proclivities, played a significant role in the judgment of her team of reviewers, Haarsma denies the obvious and, in so doing, reverses some of the progress that her reviewers had made in clarifying the real issues that separate our two groups.

Notes

1. Deborah Haarsma "Reviewing 'Darwin's Doubt': Conclusion," *The BioLogos Forum*, January 19, 2015, http://biologos.org/blog/reviewing-darwins-doubt-conclusion.

2. Ibid.

3. Darrel Falk, "Further Thoughts on 'Darwin's Doubt' After Reading Bishop's Review (Reviewing 'Darwin's Doubt': Darrel Falk, Part 2), *The BioLogos Forum*, September 11, 2014, http://biologos.org/blog/thoughts-on-darwins-doubt-reviewing-darwins-doubt-darrel-falk-part-2.

4. Ibid.

5. Ralph Stearley, "The Cambrian Explosion: How Much Bang for the Buck?," *Perspectives on Science and the Christian Faith* 65 (December 2013): 255, http://www.asa3.org/ASA/PSCF/2013/PSCF12-13Stearley.pdf. Emphasis added.

6. Douglas Erwin and James Valentine, *The Cambrian Explosion: The Construction of Animal Biodiversity* (Greenwood Village, CO: Roberts and Company, 2013), 330.

7. Stearley, "The Cambrian Explosion," 252.

8. Erwin and Valentine, *The Cambrian Explosion*, 330 (emphasis added). See also: Douglas H. Erwin and Eric H. Davidson, "The evolution of hierarchical gene regulatory networks," *Nature Reviews Genetics* 10 (February 2009): 141–148.

9. Ibid., 9.

10. As Erwin elsewhere notes, "The crucial difference between the developmental events of the Cambrian and subsequent events is that the former involved the establishment of these developmental patterns, not their modification." (Douglas H. Erwin, "Early Introduction of Major Morphological Innovations," *Acta Palaeontologica Polonica* 38 (1994): 288. In another place he writes: "There is every indication that the range of morphological innovation possible in the early Cambrian is simply not possible today." (Douglas H. Erwin, "The Origin of Bodyplans," *American Zoologist*, 39 (1999): 617–629.

11. I do acknowledge, in deference to one point made by Haarsma, that Ralph Stearley did take significant scientific issue with me about the duration of the Cambrian explosion, but my colleagues and I have replied to similar critiques. (See Chapters 6, 13, and 14 in the present volume.) In any case, Stearley's attempt to extend the time available to the evolutionary process—by defining the Cambrian explosion to include discrete events in the earlier Cambrian (such as the appearance of the small shelly fossils) or discrete events in the late Precambrian (such as the Ediacaran radiation)—does not significantly diminish the difficulty of accounting for the amount of morphological novelty that arises so abruptly in the middle Cambrian, in particular in the crucial Tommotian and Atdabanian stages of the middle Cambrian (part of what are also called Cambrian stages 2 and 3). I follow Erwin, Valentine, and other Cambrian experts in dating the duration of the Cambrian explosion as a whole to about 10 million years. But in *Darwin's Doubt* I also show, based upon separate analyses by Erwin and MIT geochronologist Samuel Bowring, that 13–16 new animal phyla arose abruptly within just a 5–6 million year window of the middle Cambrian. Including earlier discrete paleontological events (as Stearley does) within the designation "Cambrian explosion" does nothing to explain how the novel forms of animal life (and the biological information necessary to produce them) arose in such a narrow window of geological time.

12. Darrel Falk, "Thoughts on 'Darwin's Doubt' (Reviewing 'Darwin's Doubt': Darrel Falk, Part 1), *The BioLogos Forum*, Sept. 9, 2014, http://biologos.org/blog/thoughts-on-darwins-doubt-reviewing-darwins-doubt-darrel-falk-part-1.

13. It may be that Ralph Stearley betrays a commitment to methodological naturalism in another way. He claims that the Cambrian explosion encompasses 25 million years or more. In so doing, he treats numerous discrete paleontological events as part of one unitary, continuous evolutionary event. He also repeatedly claims that nothing about the Cambrian fossil record "negate[s] a genealogical organization to life." By this, he clearly means to affirm the Darwinian universal tree of life with its depiction of continuous (rather than discrete or discontinuous) morphological change as the best representation the history of animal life. But on what basis does he affirm this? In his companion review of Erwin and Valentine's book, he acknowledges that they describe the discontinuous nature of the Precambrian-Cambrian fossil record and that those "discontinuities" remain mysterious from an evolutionary point of view. Certainly, the series of discrete events in the history of life—the Ediacaran radiation, the appearance of the small shelly fossils, the first appearance of most animal body plans in the middle Cambrian—that Stearley wants to fuse together and call the Cambrian explosion are not documented as a series of continuous morphological transformations in the fossil record. And, contrary to what he claims in his review, I provide an extensive discussion (an entire chapter in fact, not just a few pages) of why most Cambrian paleontologists don't regard the Ediacaran fauna as ancestral to the main groups of animals that arise in the middle Cambrian. (In brief, the Ediacaran fauna lack discernable morphological or anatomical affinities with those later Cambrian forms and, in the view of many, were likely not even animals.) In addition, Stearley acknowledges that attempts to reconstruct the animal tree of life have resulted in conflicting trees depending upon which molecules or anatomical characters are analyzed (though he dismisses these anomalies as being inconsequential and entirely expected). So if neither the paleontological, nor the genetic and anatomical evidence, unambiguously support the monophyletic picture of the history of animal life, why not take the discontinuity of the fossil record at face value and at least consider a polyphyletic picture instead? One reason not to consider that possibility is that such a picture of the history of life would imply radical breaks in the continuity of natural evolutionary processes. In other words, that picture of the history of life implicitly challenges a seamless naturalistic unfolding of animal life. But if the evidence doesn't clearly support such a picture, what does? Could Stearley's confidence in "the genealogical organization of life" again reflect tacit extra-evidential commitments to portraying the history of life as the outworking of purely naturalistic processes—i.e., a commitment to methodological naturalism? I think it's a fair question.

14. Robert Bishop, "Meyer's Inference to Intelligent Design as the Best Explanation (Reviewing 'Darwin's Doubt': Robert Bishop, Part 3)," *The BioLogos Forum*, Sept. 8, 2014, http://biologos.org/blog/meyers-inference-to-intelligent-design-as-the-best-explanation-reviewing-da.

15. "How is BioLogos different from Evolutionism, Intelligent Design, and Creationism?," The BioLogos Foundation, http://biologos.org/questions/biologos-id-creationism, emphasis added.

16. For some other examples of BioLogos authors endorsing the principle of methodological naturalism, even sometimes using the term "methodological naturalism" itself, see Robert Bishop, "Meyer's Inference to Intelligent Design"; Ted Davis, "Searching for Motivated Belief: Understanding John Polkinghorne, Part 2," The BioLogos Forum, March 14, 2013, http://biologos.org/blog/searching-for-motivated-belief-understanding-john-polkinghorne-part-two/P0; Darrel Falk, "Signature in the Cell," The BioLogos Forum, Dec. 28, 2009, http://biologos.org/blog/signature-in-the-cell/P60; Mark H. Mann, "Let's Not Surrender Science to the Secular World! Part 5," The BioLogos Forum, Feb. 6, 2012, http://biologos.org/blog/lets-not-surrender-science-to-the-secular-world-part-5; Mark Sprinkle, "Teaching the Whole Controversy," The BioLogos Forum, April 22, 2012, http://biologos.org/blog/teaching-the-whole-controversy.

41.

AMONG THEISTIC EVOLUTIONISTS, NO CONSENSUS

Casey Luskin

IN REVIEWING *DARWIN'S DOUBT,* EVEN ALMOST A YEAR AND A HALF since it came out, theistic evolutionists could not seem to agree on what Stephen Meyer got wrong. As David Klinghoffer writes in Chapter 36, when Darrel Falk reviewed Meyer's book, he agreed that Stephen Meyer is right to point out that leading evolutionary theorists are in the process of rethinking important neo-Darwinian claims. Most fundamentally, they are reconsidering whether the standard model can account for large-scale macroevolutionary change. In noting this, Falk (a biologist) explicitly disagreed with a critical review of Meyer's book posted at BioLogos by Wheaton College philosopher Robert Bishop, who claimed that the neo-Darwinian paradigm was doing just fine.

BioLogos subsequently posted the text of a speech by Alister Mc-Grath, framed in the headline so as to suggest that Meyer was guilty of making a "God of the gaps" argument. (See Chapter 38.) Subsequently Bishop co-authored *another* critical review of *Darwin's Doubt* along with Meyer's *Signature in the Cell,* this article appearing in *Christianity Today's* review journal *Books & Culture.*[1]

Bishop's second critique is noteworthy for its concession that Meyer does not in fact make a "God of the gaps" argument. He also acknowledges that Meyer's is not an "argument from ignorance." Along with Wheaton College philosopher Robert O'Connor, Bishop writes that "Meyer deftly dispatches... the misconception that [intelligent design] engages in crude god-of-the-gaps reasoning or presents a simplistic argument from ignorance."[2]

That basically defeats the previous attempt by BioLogos to portray *Darwin's Doubt* as a gaps-based argument. Bishop and O'Connor also deserve credit for avoiding some common traps among critics of Meyer's work. Beyond that, unfortunately, their review is marred by serious errors.

They accuse Meyer of "begging the very question at hand," that is, whether there might be other unknown material causes that could produce complex and specified information (CSI) in life. They write:

> [T]his phrase, "only one known cause," is crucially ambiguous. It might mean that, among all the possible causes, there is only one that we have good reason to believe is capable of producing specified complexity. This point, however, poses (could there be others?) rather than answers the question.[3]

By appealing to unknown causes to block the design inference, they effectively commit a materialism-of-the-gaps fallacy. That is, they assume that material causes will be discovered to explain all things and thus that we can never infer design.

But why are Bishop and O'Connor so concerned about unknown causes in the first place? It seems to be because they misread Meyer as saying that "we have positive knowledge that no other causes are adequate." In other words, they think Meyer is affirming that no other possible causes, known or unknown, can explain life's high CSI. But that's not at all what Meyer says. In fact, in arguing his case, Meyer nearly always inserts the word "known" before "cause." For one of many examples:

> But philosophers of science have insisted that assessments of explanatory power lead to conclusive inferences only when there is just one known cause for the effect or evidence in question.[4]

Here's another:

> Only if the Cambrian event and animals exhibit features for which intelligent design is the only known cause may a historical scientist make a decisive inference to a past intelligent cause.[5]

Indeed, Bishop and O'Connor's review includes multiple citations from Meyer where he inserts "known" before "cause," yet they misrepresent Meyer's argument as saying the opposite. Meyer doesn't claim to have exhaustive knowledge of all possible causes, even those presently unknown. He only claims to refute *known* material causes.

Moving along, Bishop and O'Connor claim that Meyer offers "very little substantive support for mind having unique causal properties" other than the fact that mind is "immaterial." Again, one could cite many passages from Meyer's writings that clearly show their characterization is wrong. This time, let's take an example from *Signature in the Cell*:

> [O]ur uniform experience affirms that specified information— whether inscribed in hieroglyphics, written in a book, encoded in a radio signal, or produced in a simulation experiment—always arises from an intelligent source, from a mind and not a strictly material process. So the discovery of the specified digital information in the DNA molecule provides strong grounds for inferring that intelligence played a role in the origin of DNA. Indeed, whenever we find specified information and we know the causal story of how that information arose, we always find that it arose from an intelligent source. It follows that the best, most causally adequate explanation for the origin of the specified, digitally encoded information in DNA is that it too had an intelligent source. Intelligent design best explains the DNA enigma.[6]

Clearly Meyer provides strong positive reasons to understand why intelligence, a goal-directed cause, can produce the kind of functional digital information we see in DNA. This argument is not grounded solely in the fact that intelligence is "immaterial," but primarily in the fact that intelligent agents are able to think with an end-goal in mind and quickly find unlikely solutions to complex problems.

But we haven't yet addressed what I believe to be the most off-base critique from Bishop and O'Connor. They object to Meyer's arguing that life has properties "like computers," further saying "talk of 'genetic codes' and 'information processing' with respect to the origin of life or the nucleus can be very limiting if not misleading." This is a surprising criticism. Many leading scientists have acknowledged that DNA contains functional digital information, just as computer codes and sections of written text do. Bill Gates observes, "Human DNA is like a computer program but far, far more advanced than any software we've ever created."[7] Craig Venter says that "life is a DNA software system,"[8] that "DNA is the software of life,"[9] containing "digital information" or "digital code," and that the cell is a "biological machine" full of "protein robots."[10]

Richard Dawkins has written that "[t]he machine code of the genes is uncannily computer-like."[11] Even Francis Collins—perhaps the most famous and influential theistic evolutionist of them all—notes, "DNA is something like the hard drive on your computer," containing "programming."[12]

Many scientists similarly acknowledge that DNA uses an information-rich "code"—a digital one in fact. As a *Nature* paper titled "The digital code of DNA" explains: "DNA can accommodate almost any sequence of base pairs—any combination of the bases adenine (A), cytosine (C), guanine (G) and thymine (T)—and hence any digital message or information."[13] MIT engineer Seth Lloyd elaborates on how DNA carries digital information:

> DNA is very digital. There are four possible base pairs per site, two bits per site, three and a half billion sites, seven billion bits of information in human DNA. There's a very recognizable digital code of the kind that electrical engineers rediscovered in the 1950s that maps the codes for sequences of DNA onto expressions of proteins.[14]

Here we have leading scientists in agreement that DNA uses a code that undergoes computer-like information processing. Indeed, just about every single molecular biologist on earth would concede that genetic codes and computer-like information processing are at the heart of life. Yet when Stephen Meyer observes that life involves "genetic codes" and "information processing," he's accused of being "misleading." The double standards that ID proponents face—even from theistic evolutionists writing in the pages of Christian journals—are one of the most unfortunate peculiarities of the evolution debate.

Bishop and O'Connor seem to miss another point: Meyer never simply equates life with computers, although the computer-like properties of life indeed require an intelligent cause. Yet we note that where life's properties aren't exactly like computers, they are typically *more* complex than human technology, making the need for design even more apparent.

Finally, Bishop and O'Connor provide a helpful clarification, arguing that "mechanisms such as mutation and natural selection are not, in fact, 'wholly blind and undirected.'" They're entitled to believe that God guided the processes that created life in such a way that they appear unguided. I wonder, though, if these reviewers can provide an explanation for how God might guide an unguided process. Certainly, they should not object if few find their position compelling. The notion that mutation and selection really aren't blind and undirected is a faith-statement for which they can provide no supporting evidence.

In fact, they admit this, stating: "On the evolutionary creationist account, the work is signed using invisible ink." This is an important clarification. These two theistic evolutionists believe we cannot empirically detect God's handiwork, an idea at variance with the Apostle Paul's statement that God is "clearly seen" in nature (Romans 1:20). Thus, while their review opens with the statement that "All Christians affirm design because the entire universe is the creative work of God," they have no empirical or scientific way to back that up.

Crucially, Bishop and O'Connor's review never addresses the central question of both *Darwin's Doubt* and *Signature in the Cell*: What material causes can produce life's information-rich systems? They provide no answer, but the theory of intelligent design does. And ID's answer gives people what "evolutionary creationism" cannot: scientifically sound reasons to believe that life is the result of intelligent design.

Notes

1. Robert Bishop and Robert O'Connor, "Doubting the Signature: Stephen Meyer's case for intelligent design," *Books & Culture*, October 17, 2014, http://www. booksandculture.com/articles/2014/novdec/doubting-signature.html?paging=off.

2. Ibid.

3. Ibid.

4. Stephen C. Meyer, *Darwin's Doubt: The Explosive Origin of Animal Life and the Case for Intelligent Design* (New York: HarperOne, 2014), 349.

5. Ibid., 352.

6. Stephen Meyer, *Signature in the Cell: DNA and the Evidence for Intelligent Design* (New York: HarperOne, 2009), 347.

7. Bill Gates, Nathan Myhrvold, and Peter Rinearson, *The Road Ahead: Completely Revised and Up-To-Date* (New York: Penguin Books, 1996), 228.

8. J. Craig Venter, "The Big Idea: Craig Venter on the Future of Life," *The Daily Beast*, October 25, 2013, http://www.thedailybeast.com/articles/2013/10/25/the-big-idea-craig-venter-the-future-of-life.html.

9. J. Craig Venter, *Life at the Speed of Light: From the Double Helix to the Dawn of Digital Life* (New York: Viking, 2013), 7.

10. Casey Luskin, "Craig Venter in Seattle: 'Life Is a DNA Software System,'" *Evolution News & Views*, October 24, 2013, http://www.evolutionnews.org/2013/10/craig_ venter_in078301.html.

11. Richard Dawkins, *River Out of Eden: A Darwinian View of Life* (New York: Basic Books, 1995), 17.

12. Francis Collins, *The Language of God: A Scientist Presents Evidence for Belief* (New York: Free Press, 2006), 91.

13. Leroy Hood and David Galas, "The digital code of DNA," *Nature* 421 (January 23, 2003): 444-448, 444, http://www.nature.com/nature/journal/v421/n6921/full/nature01410.html.

14. Seth Lloyd, "Life: What a Concept!," *Edge*, August 27, 2007, http://www.edge.org/documents/life/lloyd_index.html.

42.

DENYING THE SIGNATURE:

A RESPONSE TO BISHOP

AND O'CONNOR

Stephen C. Meyer

B Y NOW READERS WILL KNOW THE CENTRAL ARGUMENT OF *Darwin's Doubt*, namely, that the functional biological information necessary to build the Cambrian animals is best explained by the activity of a designing intelligence, rather than by an undirected (i.e., materialistic) evolutionary process. To date, most reviews of *Darwin's Doubt* have not attempted to refute this argument, but have instead disputed the book's secondary argument about the discontinuity of the Cambrian-Precambrian fossil record (using cladistics, for example—see Chapters 4 and 7–9 in the present book); or they have contested more minor factual claims (such as my characterization of the brevity of the Cambrian explosion—see Chapter 6). Charles Marshall's review (see Chapters 10–14) stands as a solitary, but welcome, exception to this generalization.

In a cleverly titled joint review of *Darwin's Doubt* and *Signature in the Cell* ("Doubting the Signature," November–December, 2014, *Books & Culture*) philosophers Robert Bishop and Robert O'Connor also attempt to refute the central information-based argument for intelligent design of these books.[1] Nevertheless, they do not provide a *scientific* refutation to the main thesis of either book. In particular, they do not offer a better (or even an alternative) causal explanation for the vast amounts of novel genetic (and epigenetic) information that arises in the Cambrian period—i.e., the subject of *Darwin's Doubt*. Nor do they provide an alternative explanation for the origin of the information necessary to produce the first living cell—the subject of *Signature in the Cell*. Instead, they lodge various philosophical objections to my argument for design. In particular, they either dispute (a) the validity of the argument for intelligent design as an explanation for the origin of biological information, or they dispute (b) my characterization of what needs to be explained. In this chapter, I will examine each of these different types of critique.

Disputing the Validity of the Argument for Design

BISHOP AND O'Connor acknowledge that *Darwin's Doubt* and *Signature in the Cell* "deftly dispatch" the "misconception that [ID] engages in crude god-of-the-gaps reasoning"—a misconception that scholars associated with the BioLogos Foundation such as Bishop and Alistair McGrath have frequently promulgated (most recently in the multi-part review of *Darwin's Doubt* on the BioLogos website).[2]

Oddly, though Bishop and O'Connor concede that *Darwin's Doubt* and *Signature in the Cell* do not make arguments from ignorance (or commit the "god-of-the-gaps" fallacy), they critique the books as if they did! True, they use slightly different terminology in developing their objection. Instead of saying my case for intelligent design is based on ignorance or gaps in knowledge, they claim the books are guilty of "begging the question" about what we may learn in the future. But the substance of the objection is the same. I argue that intelligent design provides the best explanation for the origin of the biological information necessary to produce the anatomical novelty and complexity that arises in the history of life. My argument begs the question, in their view, because some as-yet-unknown cause—one of which we are presently ignorant—may eventually be discovered that will explain the origin of biological information.

Of course, I readily concede this as a possibility in the books: Clearly, we do not know anything about causes that we have yet to discover or observe. Nevertheless, Bishop and O'Connor claim that *Darwin's Doubt* and *Signature in the Cell* argue that "we have positive knowledge that *no other* causes" could in principle explain the origin of life's information-rich systems.[3] Yet, neither of my books anywhere claims exhaustive knowledge of the causal powers of all *possible* material processes, including unknown or not-as-yet-postulated causes. The books only claim to demonstrate the inadequacy of *known* (or postulated) materialistic processes and the adequacy of intelligent agency based upon uniform and repeated human experience to this point. That is why I repeatedly insert the word "known" before "cause" in my arguments. I also claim to infer intelligent design as the best explanation based upon our present knowledge, rather than trying to prove the theory of intelligent design with apodictic certainty.

As I note in the books, critics may choose to characterize this argument as an argument from ignorance if they like (or "begging the question" about what we may discover in the future, as Bishop and O'Connor do), but all scientific arguments, especially competing evolutionary arguments about the causes of past events in the history of life, have a similar logical structure and are subject to similar limitations. Indeed, it is an unavoidable aspect of the human condition that we can make no claims about the adequacy of causal processes that we have neither observed nor imagined. Scientists can only make inferences based upon our past and current knowledge of the causal powers of various entities and processes. Alas, we have no other kind of scientific knowledge.

Moreover, my arguments do not have the logical structure of a fallacious argument from ignorance. In an explanatory context, arguments from ignorance have the form:

Premise One: Cause X cannot produce or explain evidence E.

Conclusion: Therefore, cause Y produced or explains E.

Critics of intelligent design commonly claim that the argument for intelligent design takes this form as well. Michael Shermer, for example, insists that "intelligent design... argues that life is too specifically complex... to have evolved by natural forces. Therefore, life must have been created by... an intelligent designer."[4] In short, critics claim that ID proponents argue as follows:

Premise One: Material causes cannot produce or explain functional (or specified) information.

Conclusion: Therefore, an intelligent cause produced functional (or specified) biological information.

If proponents of intelligent design were arguing in the preceding manner, we would be guilty of arguing from ignorance. But the arguments for intelligent design in *Signature in the Cell* and *Darwin's Doubt* do not have this form. Instead, they assume the following form:

Premise One: Despite a thorough search, no material causes have been discovered with the demonstrated capacity to produce the functional (or specified) information present in living systems.

Premise Two: Intelligent causes *have* demonstrated the power to produce large amounts of functional (or specified) information.

Conclusion: Intelligent design constitutes the best, most causally adequate, explanation for the functional (or specified) information in the cell.

As one can see, in addition to a premise about how material causes lack demonstrated causal adequacy, my arguments for intelligent design as a best explanation also affirm (and demonstrate) the causal adequacy of an alternative cause, namely, intelligent agency. As I explained in *Signature in the Cell*:

> We also *know* from broad and repeated experience that intelligent agents can and do produce information-rich systems: we have positive experience-based knowledge of a cause that is sufficient to generate new specified information, namely, intelligence. We are not ignorant of how information arises. We know from experience that conscious intelligent agents can create informational sequences and systems. To quote [Henry] Quastler again, "The creation of new information is habitually associated

with conscious activity." Experience teaches that whenever large amounts of specified complexity or [functional] information are present in an artifact or entity whose causal story is known, invariably creative intelligence—intelligent design—played a role in the origin of that entity. Thus, when we encounter such information in the large biological molecules needed for life, we may infer—based on our *knowledge* of established cause-and-effect relationships—that an intelligent cause operated in the past to produce the specified information necessary to the origin of life.[5]

Thus, my argument does not just demonstrate the inability of one type of cause to produce biological information and then fallaciously infer, on that basis alone, that another cause did so (i.e., without demonstrating the adequacy of the proposed alternative cause). In other words, my arguments do not fail to provide a premise offering positive evidence or reasons for preferring an alternative cause or proposition as critics claim. Instead, my arguments specifically include and justify such a premise. Bishop and O'Connor claim otherwise, stating that "Meyer offers very little substantive support for mind having unique causal properties."[6] In fact, both of my books cite numerous examples from (a) ordinary experience, (b) computer "simulations" of evolutionary processes, and (c) origin-of-life simulation experiments showing that conscious and rational agents have the causal power to generate functional or specified information. My argument for intelligent design not only includes a premise affirming the positive causal powers of an alternative cause (i.e., intelligent agency); it also justifies that premise with multiple examples of those causal powers at work. Therefore, it does not commit the informal logical fallacy of arguing from ignorance. Neither does it beg the question about what we may discover about causal processes in the future; instead, it makes no claims about such as yet unknown processes. It claims only that intelligent design provides the best explanation based upon what we know now.

It's worth noting that none of the reviews of *Darwin's Doubt* or *Signature in the Cell* have refuted (and few have even challenged) either of the two key empirical premises in my arguments for intelligent design as a best explanation—as, indeed, Bishop and O'Connor themselves have not done. For obvious reasons, critics have not disputed my claim that intelligent agents have demonstrated the power to produce functional information and information-rich processing systems. (Bishop and O'Connor merely claim—mistakenly—that I did not *justify* that assertion.) Nor, perhaps surprisingly, have critics attempted to demonstrate that standard evolutionary mechanisms can account for the origin of biological information and information processing systems. Indeed, biologist Darrel Falk, one of O'Connor and Bishop's fellow theistic evolutionists (and with Bishop a BioLogos website contributor) has graciously conceded that *Darwin's Doubt* correctly claims that the neo-Darwinian mutation/selection mechanism has failed to account for the origin of major macroevolutionary events such as the Cambrian explosion of animal life. Falk further concedes that none of the other more recently proposed models of evolutionary theory have yet succeeded in this endeavor.[7]

Secular scientific critics of the argument in my book, for their part, have typically either (a) begged the question about the origin of genetic information by assuming the existence of other unexplained sources of information in order to account for specific informational increases in the history of life;[8] or (b) simply ignored the central question posed by the books and quibbled about secondary scientific issues or philosophical matters.[9]

Though they do attempt a philosophical refutation of the main information-based argument of the books (as we have seen), Bishop and O'Connor conspicuously avoid offering, or even citing, an alternative scientific explanation for the origin of biological information during the history of life. Instead, in addition to their philosophical critique, they mainly attempt to deny my characterization of *what needs to be explained*. It is to this latter line of attack that I now turn.

Denying the Signature—Functional Information as the *Explanandum*

Philosophers of science analyzing scientific arguments make a clear distinction between what needs to be explained (the relevant facts in question) and the competing explanations of those facts. They call the former the *explanandum* and the latter the *explanans*. Bishop and O'Connor do not offer a competing explanation (another *explanans*) for the origin of biological information. Instead, they dispute my characterization of what needs to be explained (the *explanandum*). They do so in several ways.

First, they question my characterization of DNA and RNA as molecules rich in functional digital information and my characterization of the gene expression system as an "information processing system"—in so doing, presumably raising questions about the need to explain the origin of these features of living systems. Specifically, Bishop and O'Connor assert that "talk of 'genetic codes' and 'information processing' with respect to the origin of life… can be very limited if not misleading." They argue that "abstracted notions of programs and processing seem inadequate to capture the exquisite precision and reliability of these processes." In order to describe the process of protein synthesis more accurately, they argue that I should abandon an "information processing metaphor."

Bishop and O'Connor are correct that, if not carefully defined, the term information can be misleading and lead to equivocation. But both of my books not only acknowledge that failing to distinguish different types of information can lead to confusion; they take great pains to avoid such confusion. In particular, both books carefully define the type of information that I argue reliably indicates the activity of an intelligent agent (*functional* or *specified* information, also known as *specified complexity*) and distinguish it from a type of information that does not, namely, Shannon information (or mere *complexity*)—in the latter case, information that may not perform a function. I also distinguish functional information generally from a special type of functional information (semantic information) in which *meaning* is conveyed to, and perceived by, conscious agents. (See *Signature in the Cell*, Chapter 4 and *Darwin's Doubt* Chapter 8, for definitions.)

In so doing, I made clear that DNA contains *functional* information but definitely *not* semantic information. Bishop and O'Connor completely ignore this crucial discussion in their review and, consequently, express unfounded worries about the use of the term information as a "metaphor" in biology. Indeed, had I implied that the information in DNA conveyed semantic meaning, my description would have been inaccurate—and, at best, metaphorical. Nevertheless, both books clearly state that DNA contains functional or specified information and argue (based upon our uniform and repeated experience) that such information, as opposed to Shannon information, reliably indicates the activity of a designing intelligence.

As my colleague Casey Luskin establishes with extensive citations in the previous chapter, no serious biologist post-Watson and Crick has denied that DNA and RNA contain functional information expressed in a digital form—information that directs the construction of functional proteins (and editing of RNA molecules). Thus, *contra* Bishop and O'Connor, my characterization of DNA and RNA as molecules that store functional or specified information is not even remotely controversial within mainstream biology.

Nor is my judgment controversial that the gene expression system (the system by which proteins are synthesized in accord with the information stored on the DNA molecule) constitutes an information processing system. That is what the network of proteins and RNA molecules involved in the gene-expression system do: They process (that is copy, translate, and express) the information stored within the DNA molecule. The information processing systems present in the cell may well be much more precise than those that human computer engineers have designed, but that does not mean that describing the gene expression system as an information processing system is inaccurate. Describing the gene expression system as an information processing system is not to employ a metaphor. It is to describe what the system does—again, to process (or express) genetic information.

Bishop and O'Connor's second objection to my characterization of what needs to be explained is that I have "presuppos[ed] an engineering picture of design." Instead, they think I should have described protein synthesis as a "teleological process." As they put it:

> Given the length of time over which developmental processes
> stretch, or the length of time over which self-replicating molecule
> must have formed in a pre-biotic environment, the abstracted no-
> tion of programs and processing seem inadequate to capture the
> exquisite precision and reliability of these processes.[10]

They also argue that describing the process of protein synthesis as an
information processing system implies "rigidly deterministic" processes,
rather than a teleological process, at work inside the cell. And they regard a teleological description of this process as "more effective and reliable as a picture of how the nucleus' processes work so well over such extended periods of time in the face of myriad contingencies." They further
insist that "the more basic self-replicating molecular processes sought by
origin of life researchers would also be goal-oriented," which they think
is "why so many biologist have continued using teleological vocabulary
and explanation in genetics."

Here Bishop and O'Connor misrepresent my characterization of
the information processing systems at work in cells and create a false
dichotomy, among other confusions. In fact, I do characterize the information processing system of the cell in teleological language, and I also
reflect on the paradox of Darwinian biologists using "incorrigibly teleological language"[11] to describe processes they believe arose through an
undirected and purposeless process. As I wrote in *Signature in the Cell*:

> Molecular biologists have introduced a new "high-tech" teleology,
> taking expressions, often self-consciously, from communication
> theory, electrical engineering, and computer science. The vocab-
> ulary of modern molecular and cell biology includes apparently
> accurate descriptive terms that nevertheless seem laden with a
> "meta-physics of intention": "genetic code," "genetic information,"
> "transcription," "translation," "editing enzymes," "signal-trans-
> duction circuitry," "feedback loop," and "information-processing
> system." As Richard Dawkins notes, "Apart from differences in
> jargon, the pages of a molecular-biology journal might be inter-
> changed with those of a computer-engineering journal."...[Thus]
> the historian of biology Timothy Lenoir observes, "Teleological
> thinking has been steadfastly resisted by modern biology. And
> yet in nearly every area of research, biologists are hard pressed

to find language that does not impute purposiveness to living forms."[12]

As the above quotation implies, an engineering picture of life *is* a teleological picture because engineers who design complex systems, including complex information processing systems, do so *purposively*. By pitting engineering design and teleology against each other, Bishop and O'Connor create a false dichotomy. They do they same by treating *determinism* and teleology as opposites. When an engineer imposes constraints on a physical system to achieve a particular functional outcome, he has an *end in mind*—thus, he is creating a teleological process. But the end he hopes to achieve will only occur if he can count on the reliability of the laws of nature—i.e., deterministic processes. All designed objects take advantage of determinist laws in order to achieve specific outcomes starting from specific sets of constrained (by the engineer) initial conditions. Teleology and determinism are not necessarily opposites because purposive agents can harness deterministic processes to achieve their desired ends. On this point, Bishop and O'Connor seem simply confused.

Thirdly, Bishop and O'Connor object to my description of living organisms as systems in which functional information is present. They contend that my characterization betrays an "objectionable" subjective element. In order to illuminate this problem as they see it, Bishop and O'Connor first attempt to distinguish between the objective and subjective aspects of my argument. They acknowledge first that some objective facts are clear:

> Biologists agree: The structure of DNA, however contingent, serves well to produce a functional outcome. There is nothing subjective in this. In spite of the complexity inherent in the coding regions of DNA, the specific arrangement "hits a functional target." That is, from among the vast array of possibilities, a DNA sequence that renders possible or enhances the life of an organism betokens the intentional activity of intelligent agency.[13]

Somewhat surprisingly, Bishop and O'Connor sound in this passage as if they accept the heart of my argument. They concede that complex sequences in the coding regions of DNA hit a "functional target"—that is, those sequences code for functional proteins (among a vast array of possible non-functional peptide sequences) and, thus, aid in the survival of living organisms. They even sound as if they are conceding that the presence of complex sequences containing *functional* information would reliably indicate intelligent design.

So what is the problem? They claim there is no objective, scientific basis for privileging, or focusing on, "life" in my analysis and that absent the assumption that life represents "a distinguished outcome," I have no objective criteria for deciding whether DNA or other bio-macromolecules represent functional outcomes, and thus, presumably that they contain *functional* information. As they put it, "inherent in the notion of a functional outcome is the presumption that life constitutes a distinguished outcome." To them, interest in life as a significant outcome reflects an objectionable and subjective value judgment. "Since life has value—to us—we naturally insist that any means conducive to life has distinctive value. *But that's an interpretation we supply.*"[14] By contrast, they argue, "An objective observer will realize that, if life is the goal, then that arrangement [of bases in a coding sequence of DNA], however, improbable, functions magnificently. If some other outcome were the goal, however—say the more modest goal of replication—then that outcome would have no particular value."

Bishop and O'Connor repeatedly claim that my argument depends upon a subjective value judgment about the importance of life. But their claim is not quite accurate. My argument does not depend upon a judgment, whether subjective or objective, about the value of life. Instead, it simply treats life as a phenomenon in need of explanation. It presupposes, as all biologists do, based upon a whole host of observations and comparisons, that life and non-life are different modes of existence and that the nature and origin of living things, therefore, requires explication and explanation.

Bishop and O'Connor are right, of course, if what they really mean is that all such observationally based judgments in science are made by human subjects—by the scientists whose subjective interests guide scientific investigations. Scientists are, after all, human beings who make judgments about which of the things they observe in the natural world seem important or unexpected or unusual or interesting and, consequently, are worth studying. In that sense, judgments about which observations and phenomena warrant special interest, or require explanation, are indeed subjective. But all scientific endeavors are motivated by subjective *human* interest and are guided by the perceptions humans have, and the judgments and observations they make, about natural phenomena. All scientific investigations depend upon what human investigators think interesting, and thus, upon that kind of subjectivity. But this is inescapable in the practice of science for the simple reason that *it is humans interested in the natural world who do science* (and, indeed, humans showing interest in the living world who do *biological* science). As philosopher of science Del Ratzsch has quipped, "science has a serious, incurable case of the humans."[15] And one thing human scientific investigators do is try to explain phenomena that, for one reason or another, seem unusual, special, curious, or unexpected to them. For almost all biologists life is one such phenomenon, "a distinguished outcome" as Bishop and O'Connor put it.

It is also true, of course, that biologists determine whether a DNA sequence performs a function by assessing whether that string will code for a protein (or an RNA) that will in turn help an organism stay *alive*. So the criterion "helps sustain life" does ultimately underlie judgments about the functionality of information-rich sequences in DNA, RNA, and proteins.

But, so what? To deny the relevance of this criterion is to treat life as something insignificant and not in need of explanation; and no scientist, especially one interested in the origin of life, does that. In any case, neither my argument, nor the validity of science itself, depends upon insisting that our collective human interest in life is entirely objective *if* by "objective" we mean somehow independent of our own interest, judgment, observations, or perceptions. The choice about whether or not to regard life as significant and in need of explanation may well reflect a subjective (i.e., human) interest in living things, and a similar recognition or perception that living things are different than inanimate rocks or chemical compounds. But that perception only renders the concept of functional information meaningless if the distinction between life and non-life is also meaningless and, again, no scientist interested in the origin of life (on any side of the debate about it) holds that view. Bishop and O'Connor may as well object to the whole field of origin-of-life research, or the entirety of the discipline of evolutionary biology, or all of biology itself, as well as to my arguments for intelligent design, since all practitioners of those fields make the same objectionable assumptions about life as "a distinguished outcome."

Regardless, determining whether cells contain functional or specified information does not require anyone to make a judgment about the value of life, but instead only a factual judgment about whether sequences of chemicals (functioning as digital characters) build protein or RNA molecules that aid in the survival of living cells. Indeed, once one has decided to regard life as a phenomenon of interest (as all evolutionary biologists do), it is objectively true that only certain arrangements of nucleotide bases, and not others, will produce proteins that perform tasks that allow cells to stay alive—a fact that Bishop and O'Connor themselves concede.

Instead of rendering the concept of functional information meaningless, Bishop and O'Connor's observation (in essence) that humans make scientific judgments about what needs explanation only makes clear that the notion of functional information depends upon a wider context of inquiry and interest that human scientists necessarily help to define. Bishop and O'Connor themselves recognize this but regard it as problematic for my argument, asserting that the assumption that life requires special explanation begs the question in favor of the design hypothesis from the start. As they put it: "[C]an one assign a function, an intended role, to a natural phenomenon without first supposing that the broader context has a specific function? To speak of the function of particular phenomena is already to have provided an answer to this global question in favor of design."

But is this really true? Does describing a biological system—a polymerase or DNA molecule, a beak or a wing, a fin or a gill—by reference to its function bias the discussion of biological origins in favor of intelligent design? Does presupposing a distinction between a functioning organism, on the one hand, and its non-functioning remains or an inanimate object, on the other, do the same? I doubt many evolutionary biologists, all of whom accept these same distinctions and functional descriptions that I do, would accept that judgment.

To describe the functional information in a living system, and to treat it as something in need of explanation, is not to say anything about how that system *originated* one way or another. There is no *a priori* or logically necessary reason that an explanation either involving, or precluding, agency must be true simply because the description of the thing to be explained includes functional language (or simply because it presupposes that life is a "distinguished outcome"). Since 1859, Darwinism and neo-Darwinism have attempted precisely to show that the appearance of design (apparent teleology) could be explained as the result of an undirected process that merely mimics the powers of a designing intelligence. Thus, it does not follow that even if some of the functional features of a living organisms appear designed that they necessarily are designed—as our Darwinian colleagues have long insisted. Instead, it is at least logically possible that a materialistic evolutionary explanation, or some purely natural process, can account for the functional features of living organisms, including their functional digital information, without recourse to a designing intelligence. If not, what has evolutionary theory been about since 1859? Most evolutionary theorists are committed to the idea that some materialistic process with sufficient creative power to generate the complex functional features of livings systems does exist or will eventually be found.

Clearly, describing the cell as system rich in functional information, or assuming that life as a phenomenon warrants explanation and scientific interest, does not *logically* entail the conclusion of design. Instead, the conclusion of design arises from a through search for, and evaluation of, the causal powers of competing possible causes and processes and the *a posteriori* discovery based upon such an examination (which my books undertake) that only one such cause, namely, intelligent agency, has the demonstrated power to produce the key effect in question: functional digital information.

Since every evolutionary biologist believes that life represents a "distinguished outcome" in need of explanation, and that living organisms have functional features produced in part as the result of genetic information, it hardly seems question begging to make the same assumption in the process of arguing for a particular theory (intelligent design) as the best explanation of those features. All theoretical contenders must do the same. Moreover, since all known forms of life require genetic (and epigenetic) information as a condition of their existence, origin, and maintenance, leading evolutionary theorists have increasingly defined the problem facing evolutionary theory, just as I do, in functional and informational terms. As Bernd-Olaf Küppers, the distinguished origin-of-life theorist, has explained, "the problem of the origin of life is clearly basically equivalent to the problem of explaining the origin of biological information."[16]

Methodological Naturalism and Materialism-of-the-Gaps

DESPITE THEIR multi-pronged critique, O'Connor and Bishop offer no evolutionary mechanism as an explanation for the origin of the information necessary to produce novel forms of animal life. Neither do they think it necessary to defend the creative power of the natural selection/random mutation mechanism, even though many leading evolutionary theorists now question its ability to generate fundamental innovation in biological form and/or information. To Bishop and O'Connor, it is enough to affirm that God uses (or could use) the natural selection/mutation process, though, they hasten to add, He necessarily does so without leaving any trace of His handiwork behind. "On the evolutionary creationist account, the work is signed using invisible ink," they aver.

In truth, the "evolutionary creationist" account that Bishop and O'Connor articulate in their review, and that they critique me for not taking seriously enough, has no empirical content beyond neo-Darwinism—although, of course, it can be accommodated to other versions of evolutionary theory as well. For example, in his *BioLogos Forum* review of *Darwin's Doubt*, Bishop (writing solo) acknowledges the incompleteness of the neo-Darwinian mechanism, but affirms, without much elaboration or explanation, that other unspecified evolutionary mechanisms have compensated (or, at least, will eventually compensate) for any deficiencies as part of an "extended synthesis."[17]

The biological details here seem unimportant to Bishop. What *is* important to proponents of evolutionary creation (EC) or theistic evolution (TE) such as Bishop and O'Connor is affirming that God works through, and only through, secondary causes. Whether there is presently any such evolutionary process that has demonstrated the capacity to generate functional digital information or biological novelty generally matters less than affirming that some such process will eventually account for the exquisite complexity of living things. However, in expressing this confidence in the inevitable success of some naturalistic explanation, proponents of EC (or TE) commit what one might justly characterize a kind of "materialism of the gaps" fallacy. Indeed, the great virtue of Bishop and O'Connor's *Books & Culture* review is precisely the way in which it reveals their *a priori* commitment to finding naturalistic explanations for all events and features of the natural world regardless of what the evidence itself might indicate.

The discovery of digital code, hierarchically organized information processing systems, and functionally integrated complex circuits and nano-machinery would in any other realm of experience immediately and properly trigger an awareness of the prior activity of a designing intelligence—precisely because of what we know from experience about what it takes (i.e., what kind of cause is necessary) to produce such systems. But Bishop and O'Connor seem entirely unmoved by discoveries showing the existence of such informational and integrated complexity in living organisms, not because the existence of functional digital code or the nanotechnology in life is in any way in doubt, but because they have committed themselves to viewing the world as *if* it were the product of materialistic or naturalistic processes *regardless of the evidence*. (Of course, they conceptualize those processes as modes of divine action, that is, "secondary causes" in theological parlance, even when those same processes clearly lack the creative capacity necessary to explain the origin of the features of life that are attributed to them.)

Both Bishop and O'Connor are Christian defenders of the principle of "methodological naturalism"—a principle that specifies that scientists *must* explain *all* events by reference to materialistic (non-intelligent) causes whatever the evidence.[18] For this reason, their affirmation that God designed the universe, but signed His work in undetectable "invisible ink," should be taken with a grain of salt. True, the "signature" of design in nature can only be seen by those with eyes to see. But an *a priori* commitment to methodological naturalism ensures that we will never perceive (or at least acknowledge) design in nature whatever the evidence, and it codifies our innate tendency to avert our eyes from what is "clearly seen"—and from what modern biology has made increasingly clear—in "the things that are made."[19]

Notes

1. Robert Bishop and Robert O'Connor, "Doubting the Signature: Stephen Meyer's case for intelligent design," *Books & Culture*, November–December 2014, http://www.booksandculture.com/articles/2014/novdec/doubting-signature.html.

2. Alister McGrath, "Big Picture or Big Gaps? Why Natural Theology is better than Intelligent Design," *The BioLogos.org Forum*, September 15, 2014, http://biologos.org/blog/big-picture-or-big-gaps-why-natural-theology-is-better-than-intelligent-des.

3. Ibid. Emphasis added.

4. Michael Shermer, "ID Works in Mysterious Ways," *The Ottawa Citizen*, July 9, 2008, http://www.canada.com/story.html?id=711a0b47-29d5-426d-a273-a270817b000e.

5. Stephen C. Meyer, *Signature in the Cell: DNA and the Evidence for Intelligent Design* (New York: HarperOne, 2009), 376–377.

6. FOOTNOTE TEXT TO COME.

7. See Darrel Falk, "Further Thoughts on 'Darwin's Doubt' after Reading Bishop's Review (Reviewing 'Darwin's Doubt': Darrel Falk, Part 2)," *The BioLogos Forum*,

September 11, 2014, http://biologos.org/blog/thoughts-on-darwins-doubt-reviewing-darwins-doubt-darrel-falk-part-2.

8. See Charles Marshall, "When Prior Belief Trumps Scholarship," *Science* 341, no. 6152 (September 20, 2013): 1344.

9. For example see Nick Matzke, "Meyer's Hopeless Monster Part II," PandasThumb. org, June 19, 2013, http://pandasthumb.org/archives/2013/06/meyers-hopeless-2. html; John Farrell, "How Nature Works," *National Review*, September 2, 2013, https://www.nationalreview.com/nrd/articles/355862/how-nature-works.

10. Bishop and O'Connor, "Doubting the Signature."

11. *Signature in the Cell*, 22.

12. Ibid, 21–22..

13. Bishop and O'Connor, "Doubting the Signature."

14. *Signature in the Cell*, 21–22. Emphasis in original.

15. Del Ratzsch, *Nature, Design, and Science: The Status of Design in Natural Science* (Albany, NY: SUNY Press, 2001), 90.

16. Bernd-Olaf Küppers, *Information and the Origin of Life* (Cambridge: MIT Press, 1990), 170–172.

17. See my discussion of Bishop's ideas on this point in Chapter 40.

18. Regarding Robert Bishop's commitment to methodological naturalism, see the discussion by Paul Nelson in Chapter 37 of this book; regarding Robert O'Connor's commitment to methodological naturalism, see Robert C. O'Connor, "Science on Trial: Exploring the Rationality of Methodological Naturalism," *Perspectives on Science and Christian Faith* 49 (March 1997): 15-30, http://www.asa3.org/ASA/PSCF/1997/PSCF3-97OConnor.html.

19. Romans 1:20.

43.

OF MINDS AND CAUSES:

A FURTHER RESPONSE TO

BISHOP AND O'CONNOR

Stephen C. Meyer

IN THEIR *BOOKS & CULTURE* REVIEW OF *DARWIN'S DOUBT* AND *SIG-nature in the Cell*, Robert Bishop and Robert O'Connor not only claim that I provide no justification for the idea that minds have causal powers that unconscious, non-rational, material processes don't; they also claim that my "analysis assumes that... mind is fundamentally immaterial" and yet they note that I offer "very little substantive support for mind having unique causal properties *inasmuch as it immaterial*."[1] In other words, Bishop and O'Connor seem to claim that I don't justify the idea (1) that minds are *immaterial* entities distinct from physical brains; and (2) that such *immaterial* minds possess causal powers that material processes do not.

In this latter respect they are right. I do not provide a justification for what philosophers of mind call "substance dualism"—the theory of mind-brain interaction that affirms the mind is an immaterial entity distinct from the physical brain. Instead, I make clear that my case for intelligent design does not depend upon holding a particular view of the mind-body question or holding that the mind is an immaterial entity. As I explained in *Darwin's Doubt*:

> Proponents of intelligent design *may* conceive of intelligence as [ultimately] a... materialistic phenomenon, something reducible to the neurochemistry of a brain, but they may also conceive of it as part of a mental reality that is irreducible to brain chemistry or any other physical process. They may also understand and define intelligence [or mind] by reference to their own introspective experience of rational consciousness and take no particular position on the mind-brain question.[2]

It is true, as Bishop and O'Connor note, that I do in various contexts contrast the causal powers of minds or agents, on the one hand, with "strictly material processes" on the other. And by pointing this out, Bishop and O'Connor seem to be posing a philosophical dilemma for me. They seem to be suggesting, on the one hand, that because I have contrasted mind with strictly material processes, my argument presupposes that mind cannot ultimately have a materialistic basis. Thus, they assert that, "If material processes lack such causal powers [as Meyer argues], then intelligent agency cannot be material." It seems to follow for them that I cannot allow the possibility of a materialist (or physicalist) account of mind without effacing the distinction between mind and matter (or materialistic processes) that would make an inference to intelligent design significant. On the other hand, if I presuppose an immaterial conception of mind, they fault me for failing to provide a justification for such a conception (including the idea that mind conceived as an immaterial substance possesses unique causal powers).

They also argue that any potential justification for dualism would necessarily have to be philosophical, rather than scientific, in character—thus, in their view, rendering the theory of more philosophical than scientific. As they explain, "any way you look at it, what support might be available [for the idea that for mind is immaterial] must certainly be regarded as philosophical rather than scientific. At least on this side of the ledger, ID looks more like philosophy than science."

There is a straightforward way to split the horns of the dilemma that Bishop and O'Conner pose. Rather than defending substance dualism, on the one hand, or treating mental and material phenomena as indistinguishable, on the other, the case for intelligent design can be made utilizing a more philosophically minimalist or pre-theoretical conception of mind. And both my books make use of such a conception. Indeed, by making a distinction between minds and strictly material processes, I am not committing to full-blown substance dualism *as a condition of being able to make design inferences* (as my disclaimer above indicates). Instead, I assume a more philosophically minimalist (or pre-theoretic) conception of mind or intelligence that acknowledges a distinction between physical states and mental states (such as desires, thoughts, beliefs, and emotions), but one that does not insist that the distinction between these two types of phenomena necessarily derives from two different types of substances, one material and the other immaterial. Thus, my books implicitly distinguish minds from "strictly material processes" by reference to precisely those mental attributes such as "consciousness, will, deliberation, foresight and rationality" that we know minds possess as the result of introspection. As such, my argument depends only upon a distinction that nearly all people recognize as a result of their own direct awareness of mental phenomena and conscious experience.

Bishop and O'Connor acknowledge that I equate intelligent agency with "self-conscious mind in possession of thoughts, will and intentions." Hence, they seem to recognize that I define mind by reference to specific and distinctively mental properties of which we are all aware. Nevertheless, they seem to think that I need to go further and demonstrate that these properties derive from an immaterial substance in order for us to be able to detect intelligent or mental activity. Though I personally think that substance dualism has a lot of merit, I don't think that follows.

Indeed, many investigators make design inferences without having an account of the origin of the mind or the mind/brain interaction. Forensic scientists and archeologists, for example, neither presuppose substance dualism, nor reject physicalism—nor do they necessarily have any opinion on these matters—in order to infer that some objects, structures, or events are the product of a mind. Put differently, when forensic scientists or others make a design inference based on their (presumably) pre-theoretic distinction between the causal powers of agents and material processes, they do not also thereby commit themselves to any other particular view of mind-body interaction. When a mom finds a huge mess in the kitchen and infers that her kid did it (as opposed to some "natural" cause such as, perhaps, a tornado!), she can clearly do so without also justifying some substantive position in the philosophy of mind.

Similarly, a materialistically-minded scientist might infer—based upon the information-bearing properties of DNA and knowledge of the unique causal powers of intelligent agents—that a designing agent or mind of some kind played a role in the history of life. Yet that same materialist could conclude (as Richard Dawkins has allowed as a possibility) that the designing agent in question evolved, and evolved its powers of agency, by some strictly materialistic evolutionary process. I find this possibility extremely implausible, not only because I doubt that consciousness, rationality, imagination or mental *qualia* have been (or can be) explained by reference to brain chemistry, but also because this view begs crucial origins questions. If evolutionary theory has failed (as my books show) to explain the origin of the genetic information necessary to produce living systems on this planet in the first place, positing that life—and/or complex conscious life—first evolved somewhere else in the cosmos hardly solves that problem. Nor would the postulation of a wholly materialistic designing agent residing within the cosmos explain the origin of the fine-tuning of the universe itself. Clearly, no such immanent agent within the cosmos can account for the design parameters built into the very fabric of physical laws and the universe itself.

Nevertheless, I do not need to foreclose or reject the possibility of such a designing agent *a priori* in order to show that meaningful design inferences can be made or that the past activity of a designing agent *of some kind* provides the best explanation for the origin of functional biological information. We don't need to know how minds came to be, or all the necessary and sufficient conditions of mental phenomena, to infer the presence or past activity of mind from evidence that we know only minds produce. Moreover, a meaningful distinction between mind and matter (or "strictly material processes") can be justified by reference to what we know from observation and introspection about the differences between minds and material processes without such a defense. Indeed, we have ample reason for thinking—and plenty of observational evidence supporting the idea—that minds have attributes that rocks, waterfalls, chemical reactions, electromagnetic forces, genetic mutations, and tectonic plates do not.

We can, of course, *theorize* (as materialists do) that ultimately some material process—perhaps involving neurochemistry—can explain how our conscious experience arises from the material substrate of the brain. Similarly, materialists can *theorize* that somehow some evolutionary mechanism initially produced the attributes we associate with minds such as consciousness, will, reason, imagination, foresight, and the like. But positing such materialistic explanations to explain the nature and origin of conscious experience and the other known capacities of minds does not efface the distinction between mind (or mental phenomena) and matter (or material processes) that we know and observe on the basis of our ordinary experience. Indeed, *it is precisely those distinctive attributes of minds, known from uniform experience and introspection, that physicalists (or epi-phenomenalists) seek to explain.* To get *any* theory of mind off the ground, including physicalist theories, the theorist assumes the same *prima facie* distinction between mental attributes and material attributes that I presuppose in my books.

For this reason, I do presuppose a distinction between material and mental phenomena without defending, and without needing to defend, the idea that the mind is necessarily an immaterial substance. And if I don't need to justify that the mind is necessarily an immaterial substance, then it follows that I also don't need to justify the claim that the mind *as an immaterial substance* has unique causal powers that material processes lack. Strictly speaking, I need only justify the assertion that minds (as we conceive of them based on our direct pre-theoretic introspective and observational experience) have causal powers that material objects and processes do not. And both of my books certainly do that.

Though I don't need to justify substance dualism as a condition of making design inferences, it doesn't follow that there are not good justifications either for (a) *presupposing* some form of minimalist pre-theoretic form of dualism or (b) for the philosophical position of substance dualism itself. In the first place, some form of dualism may well be a properly basic belief, justified by the universal human experience of being aware of ourselves as simple, conscious subjects or "I's" distinct from our physical bodies. Since we have a similar awareness of our mind's causal powers (and their ability to exercise "downward" causation on the physical world as opposed to being a mere epiphenomenon resting inertly atop a neurophysiological substrate) our pre-theoretic awareness of these powers may well implicitly constitute a dualist understanding of mind. Yet, it does not follow from this fact that our pre-theoretic conceptions of mind need explicit philosophical justification. Instead, assuming a (minimally dualistic) conception of mind may well be a properly basic belief.

Indeed, virtually everyone accepts the belief that their minds have causal powers, including powers that material objects and process do not. Moreover, even those few materialist scientists or philosophers who deny this belief in their explicit philosophical or scientific statements betray a commitment to it in many ways as they go about their daily lives. Materialists cannot live consistently with their own denial of the causal powers of their own minds. Instead, their actions betray their belief in those powers. In addition, virtually no one gives arguments for—or, more importantly, feels the need to give arguments for—their belief in the causal powers of their own mind. And almost no one (save for a few ideologically-zealous physicalist philosophers) thinks there are defeaters for this belief. For all these reasons, it seems the common belief that our minds have causal powers, including causal powers that material objects and processes do not, seems to qualify as properly basic.

In any case, there are also good explicit scientific and philosophical arguments justifying substance dualism as a theory of mind-body interaction. See, for example, *The Mysterious Matter of Mind* by Arthur Custance, which summarizes the many neurophysiological experiments that led neuro-scientists such as Wilder Penfield and Sir John Eccles to adopt a "dualist interactionist" view of mind and brain.[3] See also Angus Menuge's *Agents Under Fire* for a good philosophical defense of substance dualism.[4] Just as there there are good philosophical arguments showing that a minimalist pre-theoretic form of dualism does not need justification (i.e., is properly basic), there are also good scientific and philosophical arguments justifying substance dualism as a good theory of mind-body interaction.

Nevertheless, Bishop and O'Connor think that because mind-body dualism requires a philosophical justification, intelligent design does not qualify as a scientific theory, but instead "looks more like philosophy than science." But that doesn't follow for several reasons already discussed: The case for intelligent design does not depend upon substance dualism; a more minimalist pre-theoretic form of dualism doesn't necessarily require any justification (and may be regarded as properly basic); and there are scientific, as well as philosophical, justifications for substance dualism (or the closely related position of dualist interactionism). In any case, many scientific theories—Einstein's theory of general relativity, Newton's theory of universal gravitation, and Darwin's theory of evolution, to cite just a few examples—are, arguably, based upon deeper philosophical premises, presuppositions, and concepts, which either can be, or need to be, justified by philosophical lines of argument. Despite their background in the philosophy of science, Bishop and O'Connor seem to assume the ability to make strict demarcations between science and philosophy in a way that philosophers of science have now almost universally repudiated (for reasons that I explain in both my books). Besides, as I argue in both books, what matters is not how we classify a theory, but whether a theory is true or warranted by the evidence.

One final point is worth making. Though I don't need to justify substance dualism as a theory of the mind as a condition of making design inferences in biology, I would certainly concede that offering a robust philosophical and/or scientific justification for such a theory would clearly enhance the philosophical importance of the case I make for intelligent design. If mind cannot be adequately accounted for by reference to materialistic processes, then any evidence of mind acting in the history of life would pose a more explicit challenge to the philosophy of scientific materialism than I develop in my books. Specifically, it would provide evidence of an *immaterial* agency acting in the history of life. If, in addition, there is strong evidence for the activity of a designing agent establishing the finely-tuned conditions of the universe present from its beginning, as I believe there is, then the conjunctions of these considerations would provide strong grounds for theistic belief.

Notes

1. Robert Bishop and Robert O'Connor, "Doubting the Signature: Stephen Meyer's case for intelligent design," *Books & Culture*, November–December 2014, http://www.booksandculture.com/articles/2014/novdec/doubting-signature.html.

2. Stephen C. Meyer, *Darwin's Doubt: The Explosive Origin of Animal Life and the Case for Intelligent Design* (New York: HarperOne, 2013), 394.

3. Arthur C. Custance, *The Mysterious Matter of Mind* (second online edition, 2001; originally published by Probe Ministries and Zondervan Publishing, 1980), http://www.custance.org/Library/MIND/.

4. Angus Menuge, *Agents Under Fire: Materialism and the Rationality of Science* (Lanham, MD: Rowman & Littlefield, 2004).

IX.

INDEPENDENT
CONFIRMATION OF
MEYER'S THESIS

[S]kepticism over evolution will soon be history.

HEADLINE, *HUFFINGTON POST*

Frank Eltman, "Richard Leakey: Evolution Debate
Soon Will Be History," *Huffington Post*, May 27,
2012, http://www.huffingtonpost.com/2012/05/27/
richard-leakey-evolution-debate_n_1548766.html.

44.

ERWIN AND VALENTINE:

THE CAMBRIAN ENIGMA

UNRESOLVED

Casey Luskin

Two months before *Darwin's Doubt* was published, Greg Mayer contributed a post on Jerry Coyne's blog, *Why Evolution Is True*, encouraging readers to buy a then-new (2013) book about the Cambrian explosion if they "really want to learn something about this period in the history of life."[1] He wasn't referring to the forthcoming book by the "infamous Stephen Meyer," as Greg Mayer called him.

Rather, Mayer suggested that people read a different book recently published by two paleontologists who are two of the leading mainstream scientific authorities on the Cambrian explosion, Douglas Erwin and James Valentine. The book is *The Cambrian Explosion: The Construction of Animal Biodiversity*.[2] I ordered it as soon as I learned it was available. Having read it, I wholeheartedly agree with Mayer that people should read *The Cambrian Explosion*. Anyone who does so will gain an appreciation of the magnitude of the explosion of biodiversity that appeared in the Cambrian, and also the size of the problem that it poses for evolutionary biology. This makes *The Cambrian Explosion* all the more worthwhile, because, as we'll see, the authors admit that from their vantage as evolutionary biologists, the Cambrian explosion is currently "unresolved."

As an initial compliment, I would like to note that Erwin and Valentine's book contains many elegant and beautiful color photos, illustrations, and diagrams of Cambrian fossils and animals. You can see some of these photos on the publisher's website.[3] (I can only imagine what the art budget was!) It also offers probably the most comprehensive defense of current evolutionary thinking about the origin of animals in the Cambrian.

But there is something even more interesting about *The Cambrian Explosion*. Erwin and Valentine are not proponents of intelligent design. So obviously they're not going to agree with everything Stephen Meyer writes in *Darwin's Doubt*, especially when Meyer argues for intelligent design. Nevertheless, if you read their book carefully, you will find that the authors articulate and affirm at least three core arguments that Stephen Meyer also makes in *Darwin's Doubt*. The introduction to their book includes clear statements of these points.

First, as the title suggests, *The Cambrian Explosion* acknowledges that the Cambrian explosion was a real event, and is not merely an artifact of an imperfect fossil record. The authors write:

> [A] great variety and abundance of animal fossils appear in deposits dating from a **geologically brief interval between about 530 to 520 Ma,** early in the Cambrian period. During this time, **nearly all the major living animal groups (phyla)** that have skeletons first appeared as fossils (at least one appeared earlier). Surprisingly, a number of those localities have yielded fossils that preserve details of complex organs at the tissue level, such as eyes, guts, and appendages. In addition, several groups that were entirely soft-bodied and thus could be preserved only under unusual circumstances also first appear in those faunas. Because many of those fossils represent complex groups such as vertebrates (the subgroup of the phylum Chordata to which humans belong) and arthropods, it seems likely that **all or nearly all the major phylum-level groups of living animals, including many small**

softbodied groups that we do not actually find as fossils, had appeared by the end of the early Cambrian. This geologically abrupt and spectacular record of early animal life is called the Cambrian explosion.[4]

Erwin and Valentine thus date the main pulse of the Cambrian explosion—when "all or nearly all the major phylum-level groups of living animals" appeared—to about 10 million years, consistent with the time-scale given in *Darwin's Doubt*. After going through some objections to the claim that there really was an explosion, they conclude it was a real event:

> Taken at face value, the geologically abrupt appearance of Cambrian faunas with exceptional preservation suggested the possibility that they represented a singular burst of evolution, **but the processes and mechanisms were elusive.** Although there is truth to some of the objections, **they have not diminished the magnitude or importance of the explosion... Several lines of evidence are consistent with the reality of the Cambrian explosion.**[5]

Second, as the book's subtitle suggests ("*The Construction of Animal Biodiversity*"), the book observes that explaining the Cambrian explosion requires explaining the origin of many diverse animal forms and body plans. Again, the authors write:

> The subtitle of this book, *The Construction of Animal Biodiversity*, captures a second theme: the importance of building the networks that mediate the interactions... **Increased genetic and developmental interactions were also critical to the formation of new animal body plans.** By the time a branch of advanced sponges gave rise to more complex animals, their genomes comprised genes whose products could interact with regulatory elements in a coordinated network. Network interactions were critical to the spatial and temporal patterning of gene expression, to the formation of new cell types, and to the generation of a hierarchical morphology of tissues and organs. The evolving lineages could begin to adapt to different regions within the rich mosaic of conditions they encountered across the environmental landscape, diverging and specializing to diversify into an array of body forms.[6]

I am not questioning whether Erwin and Valentine believe that animal body plans arose via unguided evolutionary processes. Obviously they do. What is important here is that they recognize that explaining the Cambrian explosion requires explaining how the vast complexity and diversity of animal forms arose.

Third, and most importantly, Erwin and Valentine observe that standard neo-Darwinian mechanisms of repeated rounds of microevolution are not sufficient to explain the explosion of life in the Cambrian. They note that "a third theme of this book is the **tension** between the nature of explanations for major evolutionary transitions in general and that of the Cambrian explosion in particular."[7]

That statement provides a good hint as to where they stand: The word "tension" is an artful way of saying that standard evolutionary mechanisms have a hard time accounting for the Cambrian explosion. Erwin and Valentine make this even more explicit later when they write:

> As geologists, we view this tension as a debate over the extent to which uniformitarian explanations can be applied to understand the Cambrian explosion. Uniformitarianism is often described as the concept, most forcefully advocated by Charles Lyell in his *Principles of Geology*, that "the present is the key to the past". Lyell argued that study of geological processes operating today provides the most scientific approach to understanding past geological events. Uniformitarianism has two components. Methodological uniformitarianism is simply the uncontroversial assumption that scientific laws are invariant through time and space. This concept is so fundamental to all sciences that it generally goes unremarked. Lyell, though, also made a further claim about substantive uniformitarianism: that the rates and processes of geological change have been invariant through time. Few of Lyell's contemporaries agreed with him. Today, geologists recognize that the rates of geological processes have varied considerably through the history of Earth and that many processes have operated in the past that may not be readily studied today.

The nature of appropriate explanations is particularly evident in the final theme of the book: the implications that the Cambrian explosion has for understanding evolution and, in particular, for the dichotomy between microevolution and macroevolution. If our theoretical notions do not explain the fossil patterns or are contradicted by them, the theory is either incorrect or is applicable only to special cases. Stephen Jay Gould employed the animals of the Burgess Shale and the early Cambrian radiation in his book *Wonderful Life* to advance his own view of evolutionary change. Gould argued persuasively for the importance of contingency—dependence on preceding events— in the history of life. Many other evolutionary biologists have also addressed issues raised by these events. One important concern has been whether the microevolutionary patterns commonly studied in modern organisms by evolutionary biologists are sufficient to understand and explain the events of the Cambrian or whether evolutionary theory needs to be expanded to include a more diverse set of macroevolutionary processes. We strongly hold to the latter position.[8]

I know that the passage above is a long one, but read it carefully. What are Erwin and Valentine saying? They make it clear, especially in the last couple of sentences, that they think "microevolutionary processes" are *not* "sufficient to understand and explain the events of the Cambrian." Indeed, they later argue that microevolutionary processes are not sufficient to explain macroevolutionary ones, stating: "the move from micro to macro forms a discontinuity."[9]

This means that they don't believe "uniformitarian explanations can be applied to understand the Cambrian explosion." Why? Because evolutionary mechanisms we observe in the present day operate at rates that are too slow to explain what took place in the Cambrian period. They are careful not to put it in such plain terms, but that is the essence of their argument. They do acknowledge that there was an "unusual period of evolutionary activity during the early and middle Cambrian."[10] And this is also a major argument that Stephen Meyer makes in *Darwin's Doubt*.

There are other statements in the book acknowledging how difficult it is to explain the Cambrian explosion through unguided evolutionary mechanisms. For example, Erwin and Valentine acknowledge that something remarkable happened during the Cambrian period, requiring rates and degrees of change greater than perhaps anywhere else in the history of life:

> Because the Cambrian explosion involved a significant number of separate lineages, achieving remarkable morphological breadth over millions of years, the Cambrian explosion can be considered an adaptive radiation only by stretching the term beyond all recognition... **the scale of morphological divergence is wholly incommensurate with that seen in other adaptive radiations.**[11]

Erwin and Valentine also argue that the Cambrian fauna evolved in a manner different from standard Darwinian processes, with few potential intermediates:

> [N]ovelty is rampant in the Ediacaran and Cambrian, but because **so few intermediate species have been preserved, we are not able to assess whether these novelties are more apparent than real.** The critical issue is the claim that evolutionary novelties may **arise from different mechanisms than adaptive change...** Morphologic evolution is commonly depicted with lineages more or less gradually diverging from their common ancestor. New features arise along the evolving lineages... Gould characterized this pattern as the "cone of increasing diversity," **but neither the Cambrian nor the living marine fauna display this pattern.**[12]

Stephen Meyer makes a very similar—though more detailed—argument in Chapters 4 and 5 of *Darwin's Doubt*. But probably the most striking statement by Erwin and Valentine comes when they concede that they lack resolved evolutionary explanations for how the diversity of the Cambrian animals arose, and why these basic body plans haven't changed since that time:

> The patterns of disparity observed during the Cambrian pose two **unresolved questions.** First, what evolutionary process produced the gaps between the morphologies of major clades? Second, why have the morphological boundaries of these body plans remained relatively stable over the past half a billion years?[13]

Don't miss the importance of this: Two of the leading scientists who study the Cambrian explosion acknowledge that the processes that produced the diverse body plans in the Cambrian are "unresolved." This is exactly why the journal *Science*, when reviewing Erwin and Valentine's book, stated:

> The Ediacaran and Cambrian periods witnessed a phase of morphological innovation in animal evolution unrivaled in metazoan history, yet **the proximate causes of this body plan revolution remain decidedly murky. The grand puzzle of the Cambrian explosion surely must rank as one of the most important outstanding mysteries in evolutionary biology.**[14]

Still, the fact that Erwin and Valentine acknowledge these key points in no way makes *The Cambrian Explosion* a substitute for Stephen Meyer's book. His book goes beyond Erwin and Valentine's in important ways. Most significantly, he describes and critiques the many post-Darwinian theories being proposed as alternatives to Darwinism by the growing number of evolutionary biologists who have become disillusioned with the neo-Darwinian account.

As a result, *Darwin's Doubt* is the most current and credible critique of neo-Darwinism available today, explaining the fundamental problems that hamper Darwinian theory as it attempts to explain the Cambrian explosion. As Meyer persuasively argues, there are many reasons to anticipate that what Erwin himself has elsewhere called a "paradigm shift"[15] in neo-Darwinism is just the beginning. The questioning of natural selection is the start of a process that will ultimately lead to a different kind of paradigm altogether: namely, intelligent design.

Notes

1. Greg Mayer, "Books on the Cambrian worth buying," *Why Evolution is True*, April 17, 2013, http://whyevolutionistrue.wordpress.com/2013/04/17/books-on-the-cambrianworth-buying/.

2. Douglas H. Erwin and James W. Valentine, *The Cambrian Explosion: The Construction of Animal Biodiversity* (Greenwood Village, CO: Roberts and Co., 2013).

3. See http://www.roberts-publishers.com/biology/the-cambrian-explosian-and-the-construction-of-animal-biodiversity.html.

4. Erwin and Valentine, *Cambrian Explosion*, 5. Emphasis added.

5. Ibid., 6. Emphasis added.

6. Ibid., 8–9. Emphasis added.

7. Ibid., 9. Emphasis added.

8. Ibid., 9–10. Emphasis added, internal citations omitted.

9. Ibid., 11.

10. Ibid., 6.

11. Ibid., 341. Emphasis added.

12. Ibid., 339–40. Emphasis added, internal citations omitted.

13. Ibid., 330. Emphasis added.

14. Christopher J. Lowe, "What Led to Metazoa's Big Bang?," *Science* 340 (June 7, 2013): 1170–71, http://www.sciencemag.org/content/340/6137/1170. Emphasis added.

15. Douglas H. Erwin, "Darwin Still Rules, but Some Biologists Dream of a Paradigm Shift," *New York Times,* June 26, 2007, http://www.nytimes.com/2007/06/26/science/26essay.html.

CONTRIBUTORS

Douglas Axe

Douglas Axe is the director of Biologic Institute. His research uses both experiments and computer simulations to examine the functional and structural constraints on the evolution of proteins and protein systems. After a Caltech PhD he held postdoctoral and research scientist positions at the University of Cambridge, the Cambridge Medical Research Council Centre, and the Babraham Institute in Cambridge. His work has been reviewed in *Nature* and featured in books, magazines and newspaper articles, including *Life's Solution* by Simon Conway Morris, and *The Edge of Evolution* by Michael Behe. He is a co-author, with Ann Gauger and Casey Luskin, of *Science and Human Origins* (Discovery Institute Press).

David Berlinski

David Berlinski is a Senior Fellow in the Discovery Institute's Center for Science and Culture. He is the author of numerous books, including *The Devil's Delusion: Atheism and Its Scientific Pretensions* (Crown Forum, 2008; Basic Books, 2009). Berlinski received his PhD in philosophy from Princeton University and was later a postdoctoral fellow in mathematics and molecular biology at Columbia University. He has taught philosophy, mathematics and English at Stanford, Rutgers, the City University of New York and the Université de Paris.

William Dembski

William Dembski is a Senior Fellow with the Center for Science and Culture at Discovery Institute, a Senior Research Scientist with the Evolutionary Informatics Lab, and one of the founders of the modern intelligent design movement. He holds a PhD in Mathematics from the University of Chicago and another PhD in Philosophy from the University of Illinois. He is the author or editor of more than twenty books, most recently *Being as Communion: A Metaphysics of Information* (Ashgate).

Ann Gauger

Ann Gauger is a senior research scientist at Biologic Institute. Her work uses molecular genetics and genomic engineering to study the origin, organization and operation of metabolic pathways. She received a BS in biology from MIT, and a PhD in developmental biology from the University of Washington, where she studied cell adhesion molecules involved in *Drosophila* embryogenesis. As a post-doctoral fellow at Harvard she cloned and characterized the *Drosophila* kinesin light chain. Her research has been published in *Nature*, *Development*, and the *Journal of Biological Chemistry*. She is a co-author, with Douglas Axe and Casey Luskin, of *Science and Human Origins* (Discovery Institute Press).

Tyler Hampton

Tyler Hampton is a student of molecular biology.

David Klinghoffer

David Klinghoffer is a Senior Fellow at Discovery Institute and the editor of *Evolution News & Views*. With Senator Joseph Lieberman, he is the co-author most recently of *The Gift of Rest: Rediscovering the Beauty of the Sabbath*. His other books include *Why the Jews Rejected Jesus*, *The Discovery of God: Abraham and the Birth of Monotheism*, and the spiritual memoir *The Lord Will Gather Me In*. He is a former literary editor of *National Review* magazine and a graduate of Brown University.

Casey Luskin

Casey Luskin was trained as a scientist and an attorney, having earned his bachelor's and master's degrees in earth sciences at the University of California at San Diego and a law degree from the University of San Diego. He has conducted scientific research at Scripps Institution for Oceanography and studied evolution extensively at both the undergraduate and graduate levels. He is Research Coordinator at the Discovery Institute and co-author of the popular curriculum *Discovering Intelligent Design: A Journey into the Scientific Evidence*. He is a co-author, with Douglas Axe and Ann Gauger, of *Science and Human Origins* (Discovery Institute Press).

Stephen C. Meyer

Stephen C. Meyer is director of the Discovery Institute's Center for Science and Culture (CSC) and a founder both of the intelligent design movement and of the CSC, intelligent design's primary intellectual and scientific headquarters. Dr. Meyer is a Cambridge University-trained philosopher of science and the author of peer-reviewed publications in technical, scientific, philosophical, and other books and journals. He is the author of *Signature in the Cell: DNA and the Evidence for Intelligent Design* and the *New York Times* bestseller *Darwin's Doubt: The Explosive Origin of Animal Life and the Case for Intelligent Design*.

Paul Nelson

Paul A. Nelson is a Fellow with Discovery Institute's Center for Science and Culture and Adjunct Professor in the MA Program in Science & Religion at Biola University. A philosopher of biology, he has been involved in the intelligent design debate internationally for over two decades. He received his PhD from the University of Chicago in the philosophy of biology and evolutionary theory, and his scholarly articles have appeared in journals such as *Biology & Philosophy*, *Zygon*, *BIO-Complexity*, and *Rhetoric and Public Affairs*, as well as in books such as *Signs of Intelligence* and *Darwin, Design, and Public Education*. He co-authored the textbook *Explore Evolution* and has appeared in several documentaries on intelligent design for Illustra Media.

INDEX

Made in the USA
Middletown, DE
27 August 2024

59767005R00230